Thin Films Technology

Practical Manual for the Laboratory Works

Thin Films Technology

Practical Manual for the Laboratory Works

Alexander Axelevitch
Holon Institute of Technology (HIT), Israel

NEW JERSEY • LONDON • SINGAPORE • BEIJING • SHANGHAI • HONG KONG • TAIPEI • CHENNAI • TOKYO

Published by

World Scientific Publishing Co. Pte. Ltd.
5 Toh Tuck Link, Singapore 596224
USA office: 27 Warren Street, Suite 401-402, Hackensack, NJ 07601
UK office: 57 Shelton Street, Covent Garden, London WC2H 9HE

Library of Congress Control Number: 2021948042

British Library Cataloguing-in-Publication Data
A catalogue record for this book is available from the British Library.

THIN FILMS TECHNOLOGY
Practical Manual for the Laboratory Works

Copyright © 2022 by World Scientific Publishing Co. Pte. Ltd.

All rights reserved. This book, or parts thereof, may not be reproduced in any form or by any means, electronic or mechanical, including photocopying, recording or any information storage and retrieval system now known or to be invented, without written permission from the publisher.

For photocopying of material in this volume, please pay a copying fee through the Copyright Clearance Center, Inc., 222 Rosewood Drive, Danvers, MA 01923, USA. In this case permission to photocopy is not required from the publisher.

ISBN 978-981-124-632-6 (hardcover)
ISBN 978-981-124-633-3 (ebook for institutions)
ISBN 978-981-124-634-0 (ebook for individuals)

For any available supplementary material, please visit
https://www.worldscientific.com/worldscibooks/10.1142/12528#t=suppl

Desk Editor: Joseph Ang

Typeset by Stallion Press
Email: enquiries@stallionpress.com

Abstract

Thin solid films are widely used in various applications: electronics, optics, mechanics. These applications are defined by specific properties of two- and other low-dimensional structures which are different from properties of the usual bulk materials. Properties of thin films are defined by preparation conditions and depend on plurality active factors. Properties of thin films should be characterized for right applications.

This book is devoted to the measurement techniques applied for thin films characterization. Particular attention is paid to the measurement of electrical and optical properties of grown thin films. The described measurement methods were used for characterization of the metal, semiconductor and dielectric thin films prepared while laboratory lessons were provided in the Holon Institute of Technology (HIT). So, the book is designed as a teaching aid, were the practical methods are accompanied by short theoretical explanations and real examples.

The book is intended for students studying the specialties related to micro- and optoelectronics, nanotechnologies, as well as for practical engineers working in the field.

Contents

Preface

My first introduction to the world of thin films took place long ago, when I was studying in the 10th grade of high school. We were studying Newton's rings and accompanying interference effects. I was amazed and impressed by this real influence of the material thickness on its properties. After graduating from the university, I came across thin films again, when I began working in a laboratory of thin films in the Electronic Technology Design Bureau. Since then, I have always returned to this interesting and endless world. My work in the Holon Institute of Technology (HIT) also began with the creation of the novel laboratory course devoted to thin films' deposition and characterization. This course was designed for undergraduate students. Over time, on the basis of this course, a lecture course dedicated to the physics of thin films' deposition methods and their applications has been developed.

The course "Thin Films and Microelectronics Laboratory" is a part of the track "Micro- and Optoelectronics" delivered through one semester after suitable general courses. So, it takes only 14 lessons during the semester. The course consists of several preliminary meetings explaining general things about microelectronics and thin films technology followed by some practical lessons. After measurements, the students must build the measured characteristics, process the data obtained from these characteristics and create the report which they should present on the seminar at the end of the semester. Evidently, the time allotted for the course is very restricted, therefore we can only deliver a limited number of experiments. These experiments usually consist of the creation of the heterojunction structures of metal-silicon and evaluation of their properties, deposition of the transparent conductive coatings, creation of the thin film thermocouples and thin film heaters. All these experiments require significant amounts of various measurement techniques and the ability to use them. Actually, the course is devoted to studying these measurement techniques and methods for exploitation, among other things.

Every novel field of science is based on its specific language first of all. So, we need to begin the study of the thin films technology field from its novel language — new descriptions and definitions. This language consists of defined and sufficient formal constructions. The required definitions and descriptions we will insert into the text sequentially as needed.

The book is organized as follows: the first part of the book represents three chapters dealing with the introduction to the field of thin film technologies, vacuum techniques, thin films' growth principles and physical vacuum deposition methods, as well as the magnetron sputtering technique. The fourth chapter is devoted to planning experiments, processing measurement results, and optimizing the thin-film deposition process. This part is necessary to build a reproducible technology with the minimum number of experiments. The second part is divided into the fourth and fifth chapters, devoted to measurements of the optical and electrical properties of deposited thin films and complex structures grown by thin-film methods. In the fifth chapter, we describe the optical properties of thin films and methods for studying them using spectrometry in various wavelength ranges. We also consider the simplest methods for determining the thickness of thin films using inexpensive laboratory instruments. In the sixth chapter of our book, we consider the electrical properties of thin films of various materials: metals, semiconductors, and dielectrics. We also describe heterojunction structures of the metal-silicon type and a transparent semiconducting layer on silicon. These structures represent different types of contacts: Rectifying and Ohmic, photoelectrical structures, MOS capacitors. Furthermore, we consider methods for measuring the electrical properties and parameters of materials using current-voltage and capacitance-voltage characteristics, as well as using dynamic hot-probe characteristics. Each part of the course will be accompanied by practical examples from our experience, which mean to make the learning more fruitful.

The main purpose of this book is to help students and young engineers understand the importance and basic principles of practical thin film technologies. This book can serve as a basic material for the study of laboratory courses related to thin film technologies, microelectronics and optoelectronics. Also, the author hopes that the information presented in this book will be useful to those who specialize in nanotechnology. These courses are intended for undergraduate and graduate students studying the specialties related to electronics, microelectronics and optoelectronics, as well as for practical engineers working in the field. The author wishes all future readers and users of this book a pleasant read.

Chapter 1

Preparation of the experiments

1.1 Introduction

Almost all the items that surround us and which we use are associated with the technology of thin films. Simple listing of applications of thin films shows that this is the real High-Tech: anti-reflection coating on glasses; optical filters and waveguides; hard coatings on shaving edges and on turbine blades; television, telephone and computer screens; solar cells, flash memory, semiconductor LEDs and lasers; lubricant and friction coatings; photo- and electro-chromic coatings; semiconductor micro-schemes, etc. See, for example, Fig. 1.1: all things presented here were prepared using thin film technologies.

The world of technology of thin films is a specific research area in which various scientific disciplines meet: Solid state physics, Molecular physics, Thermodynamics, Semiconductor physics, Plasma physics, Optics, Acoustics, Crystallography, Surface science, Chemistry, Microscopy, Research operation and experiment planning, Data processing, etc.

The main goal of a laboratory course for newcomer students is to introduce them to the field of practical knowledge in microelectronics and thin films preparation. Vacuum thermal evaporation and sputtering deposition methods will be demonstrated to students. Also, the various laboratory methods will be applied to characterize deposited thin films. The course consists of several lectures and an experimental

Fig. 1.1 Various things prepared using thin film technologies.

part. Thin films of different materials will be deposited and studied using the laboratory equipment. Basics of vacuum technique and electrical and optical properties of films and thin film devices will be studied in the learning process.

We live in a 3D world. The natural world that surrounds us contains various forms of materials, particularly crystals. For example, the large basalt crystals which grow in one of beautiful places in the north of Israel, presented in Fig. 1.2. These crystals, formed many years ago, present the naturally ordered 3D structures which are also called macrostructures. The main properties of such 3D materials are defined by atoms and molecules which are inside the bulk material; a number of surface atoms is significantly lower than the whole number of atoms.

In general, if we will decrease the thickness of a 3D material, we can obtain the 2D form. Figure 1.3 represents various materials with different dimensions. Here, the natural amethyst druse represents a 3D material. Such large single-crystalline forms require long-time growth and specific conditions. The single-crystalline silicon wafer with complex microelectronics circuitry grown on its surface

Fig. 1.2 Natural basalt crystals in the north of Israel.

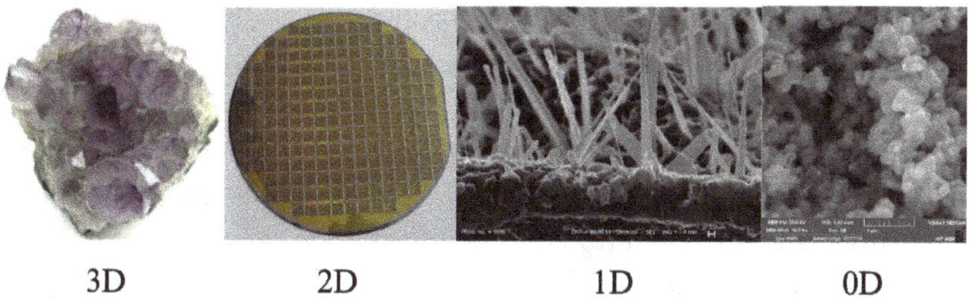

Fig. 1.3 Various types of materials.

represents the 2D material. In this case, the thickness of the silicon wafer is approximately of 300 μm, and all electron circuits on its surface together are of about 1 μm. For preparation of these electrical circuits, thin film technologies were used. The following picture represents a 1D material, the ZnO nanorods with thickness of 100–200 nm and length of 5–10 μm. These nanorods appear to have very interesting properties and may be applied for energy conversion, for example. The last photo shows the 0D material, ZnO nanoparticles. The low number of atoms in each nanoparticle allow us to call them particles without dimensions. Properties of such particles are different from the properties of bulk solid materials.

Low-dimensional materials are very important in our life. They represent such significant fields as microelectronics industry, optoelectronics, optics, mechanics, etc. The first and actual property of the low-dimensional material is the contaminations' influence on the material behavior increasing with the measurement's reducing. This is due to the number of atoms or molecules inside beginning to be comparable with the number of atoms on the boundary. Thus, the surface change, adsorption on the dangling bonds and contaminations begin to play more significant roles than in the case of 3D materials. If the thickness of a thin film is close to the crystallite's dimensions in the polycrystalline materials, the properties of the film also will be dependent on the crystallite boundary dimensions. For example, efficiency of silicon polycrystalline photovoltaic cells is lower than efficiency of single-crystalline solar cells due to scattering of the charge carriers on the boundaries between crystallites. Figure 1.4 shows the three types of silicon photovoltaic devices. Here, silicon single-crystalline solar cells are presented arranged in the panel, along with polycrystalline solar cells and thin film amorphous silicon solar cells. Evidently, efficiency of the amorphous silicon solar cells is lower among these three types. We see that all devices represent multi-layer thin film systems fabricated as the diode structures enabling the photovoltaic effect. The efficiency of the systems is defined by the level of ordering in the layers, from single-crystalline to amorphous.

Fig. 1.4 Silicon solar cells: (a) single-crystalline, (b) polycrystalline, and (c) amorphous.

Using the definition of thin solid films, we need to define the thickness of these films as a criterion. Thickness, *d*, in **one micron** ($d \leq 1\ \mu m$) is accepted as a frontier between thin and thick films. There is no doubt that this border is a stipulated value, however, it enables to divide between methods of production of the thin films and thick films. Thus, specific methods should be used for the production of thin films. The thick films have a thickness of 1–100 μm. Thicker layers are called foil or sheet. Now, it is appropriate to show the length units applied in the thin film technologies:

$$1\ \text{m} = 1 \cdot 10^3\ \text{mm} = 1 \cdot 10^6\ \text{mm} = 1 \cdot 10^9\ \text{nm} = 1 \cdot 10^{10}\ \text{Å}.$$

It is interesting to note that thickness significantly influences the properties of thin film materials. There is a so-called "size-effect" which illustrates this influence. If we denote a parameter λ as a mean free path of electrons in metal or semiconductor materials and $\gamma = d/\lambda$ will be a thickness factor, when $\gamma << 1$, the mentioned influence may be demonstrated, for example, by the following empirical relations (here, *F* denotes film and *B* denotes bulk):

- Increase of resistivity, ρ, in metal: $\rho_F/\rho_B \approx (4/3)\ [\gamma \ln(1/\gamma)]^{-1}$
- Reduced TCR, a, in metal: $\alpha_F/\alpha_B \approx [\ln(1/\gamma)]^{-1}$

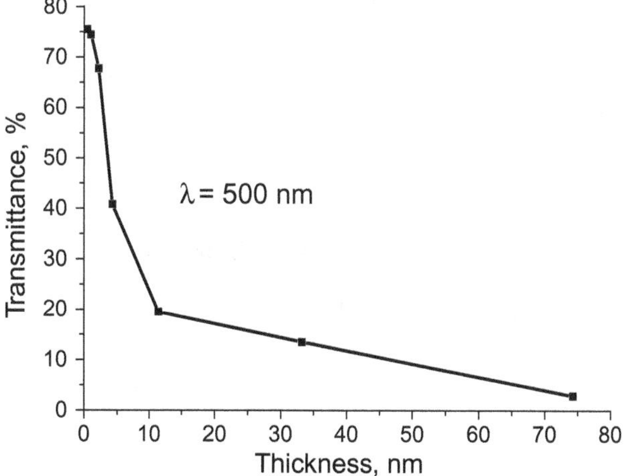

Fig. 1.5 Dependence of the thin gold films thickness on their transmittance.

- Reduced mobility, μ, in metal: $\mu_F/\mu_B \approx [\ln(1/\gamma)]^{-1}$
- Reduced thermal conductivity, K, in metal: $K_F/K_B \approx (3/4)[\gamma\ln(1/\gamma)]$
- Reduced mobility in semiconductor: $\mu_F/\mu_B \approx 1 + (1 + 1/\gamma)^{-1}$

Also, the influence of thickness on the physical properties may be illustrated by the dependence of a thickness on the transmittance on the thin gold films at the wavelength of 500 nm, as shown in Fig. 1.5.

1.2 Substrate

Each thin layer has a common element called a substrate. Thin film grows on a substrate and the substrate holds the thin film. To grow a thin film, a flux of atoms or molecules must go to the substrate. Atoms arriving on the substrate have an excess kinetic energy; these atoms begin to move on the substrate surface and find the place where they will be able to form the coupling with the surface's atoms or with the same arriving atoms. These adsorbed atoms (adatoms) can return to the volume above substrate and can form the critical dimension clusters or nucleuses which represent the beginning of the growing thin film. The critical clusters appear in the places with free bonds or on the substrate's defects that are illustrated by Fig. 1.6.

In other words, a layer always returns to the surfaces shape of the substrate. Therefore, the roughness of the substrates and their cleanness becomes very significant for growing thin films with required composition, structure and homogeneity. Roughness, crystalline structure, type of material and its composition define the thin film

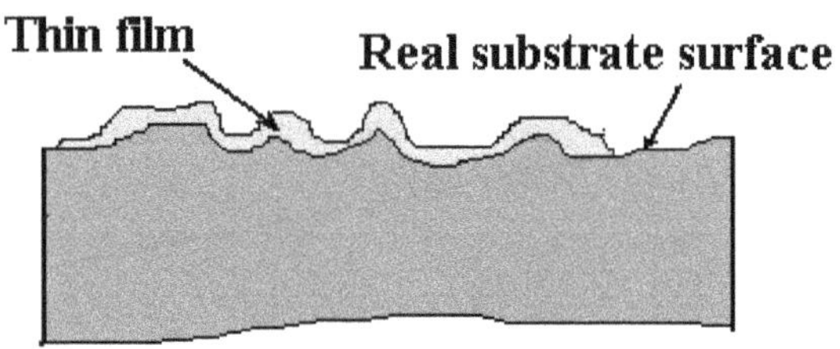

Fig. 1.6 Thin film grown on the substrate surface.

growth process. Moreover, the cleaning of the substrate before thin film deposition becomes a necessary requirement.

One of the significant technological factors is the temperature of the substrate during the thin film growth process. Figure 1.7 illustrates dependence of linear expansion of various materials on temperature.

As shown, all materials have different coefficients of thermal expansion. Moreover, such materials as glass or silicon nitride, amorphous materials, expand by the same value in all directions.

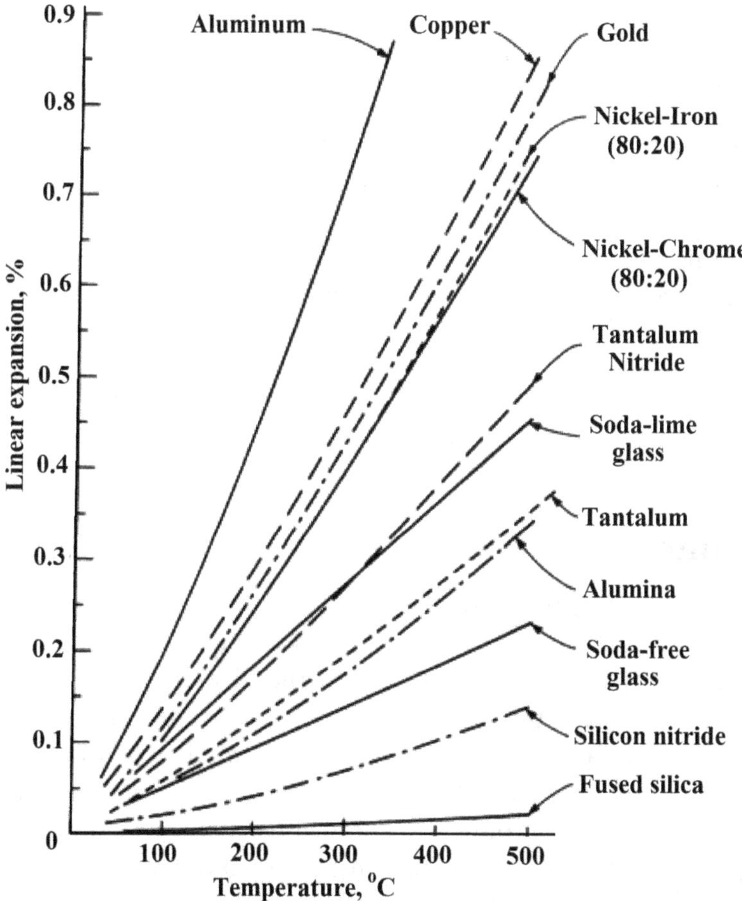

Fig. 1.7 Dependence of linear expansion on temperature for different substrates and thin films.

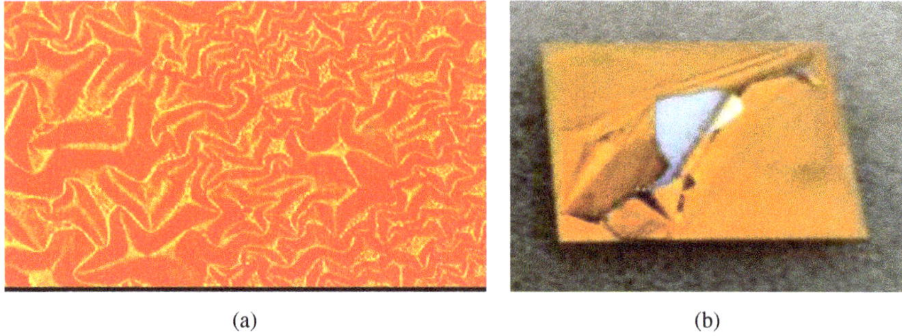

(a) (b)

Fig. 1.8 Exfoliation of deposited thin films from the substrates: (a) exfoliation of the Cu thin film from a glass substrate after heating for 400°C, (b) exfoliation and destroying of the gold layer from the installed substrate after heating up to 1000°C.

The crystalline materials such as alumina or metals have different coefficients of expansion in different directions, anisotropic materials. These effects must be considered when developing a technology of thin films preparation. Thermal behavior of thin films should be matched with the thermal behavior of substrates. In the case of no matched materials, the thin film can be exfoliated from the substrate as shown in Fig. 1.8, for example. These photographs represent thin metal films on different substrates heated for high temperature and exfoliated from the substrates.

1.3 Interaction between substrate and coating

All thin film systems produced for fabrication of different devices should have suitable lifetimes providing long, normal functioning of the devices. This requirement implies the absence of spontaneous chemical reaction between a deposit and a substrate. The possibility of spontaneous chemical reactions may be predicted using simple thermodynamic relations. As known, all chemical reactions may be divided into two big groups: Exothermal reactions, which evolve the heat, and endothermal reactions, which absorb the heat. Exothermal reaction can occur spontaneously, therefore the systems where two materials providing exothermal reactions are in contact will not be stable. Characterization of thermodynamic system may be provided using the Gibbs free energy (ΔG) of a reaction variation. A negative value for ΔG indicates that a reaction can proceed spontaneously without external inputs, while a positive value indicates that it will not. The equation for the Gibbs free energy is as follows:

$$\Delta G = \Delta H - T\Delta S \tag{1.1}$$

where ΔH is the enthalpy, T is absolute temperature and ΔS is entropy. Enthalpy (ΔH) is a measure of the actual energy that is liberated or evolved when the reaction occurs (the "heat of reaction"). If it is negative, then the reaction gives off energy, while if it is positive, the reaction requires energy. Entropy (ΔS) is a measure of the change in the possibilities for disorder in the products compared to the reactants. For example, if a solid (an ordered state) reacts with a liquid (a somewhat less ordered state) to form a gas (a highly disordered state), there is normally a large positive change in the entropy for the reaction.

A practical application for the Gibbs free energy estimation may be found, for example, in choosing a thin-film electrode metal for connecting different elements of a planar electrical circuit. The metal electrode deposited on the surface of the dielectric sublayer, usually a silicon oxide (SiO_2) layer, should be very stable to provide long, normal work of the circuit. However, sometimes the electrode's metal can react with the oxide layer by oxidation–reduction (redox) reaction with reduction of silicon from the oxide and oxidation of the electrode, thus destroying the circuit in this way. Evaluation of the possibility of redox reactions may be provided using an Ellingham diagram, shown in Fig. 1.9.

An Ellingham diagram represents a plot of free energy of oxides formation, ΔG, versus temperature. Since ΔH and ΔS are essentially constant with temperature unless a phase change occurs, the free energy versus temperature plot can be drawn as a series of straight lines, where ΔS is the slope and ΔH is the y-intercept. The slope of the line changes when any of the materials involved melt or vaporize. To predict the behavior of a metal in contact with the suitable oxide at given temperature condition, we need to compare their free energy

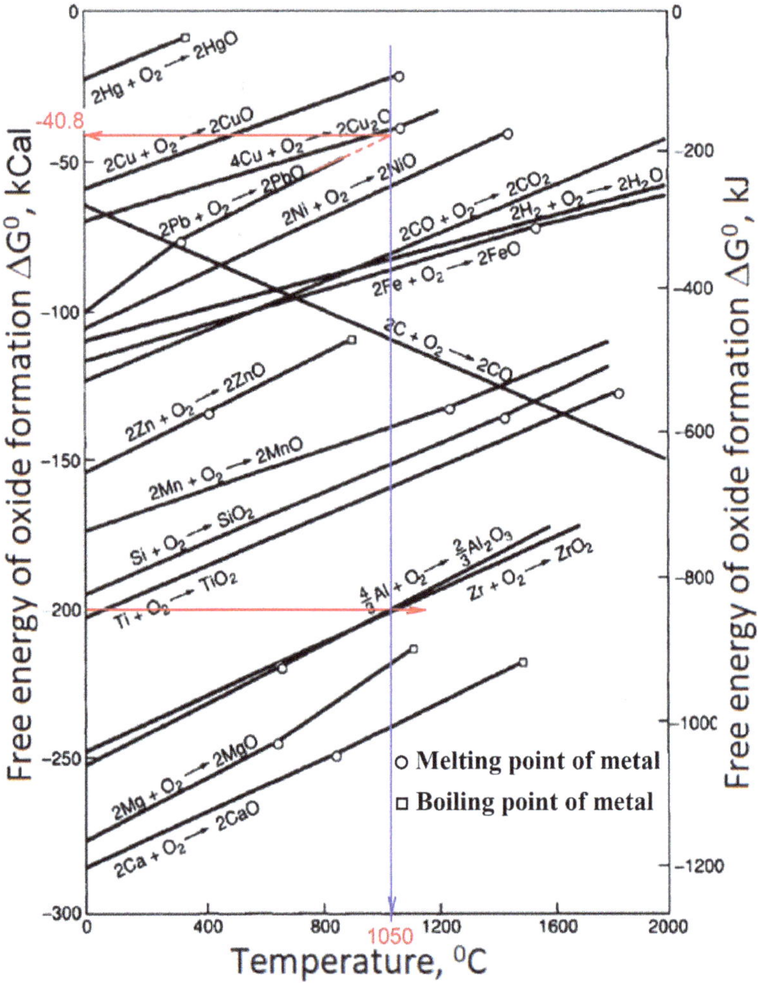

Fig. 1.9 An Ellingham diagram presents the dependence on temperature of the formation of various oxides.

of oxide formation. If the resulting energy will be positive, the projected system will be stable. Now, it is appropriate to show the energy units applied in thin films technologies:

$$1\text{J} = 10^7 \text{ erg} = 0.239 \text{ Cal} = 9.87\cdot10^{-3} \text{ lit}\cdot\text{atm} = 0.62\cdot10^{19} \text{ eV}.$$

For example, we can consider the choice of a suitable metal from among aluminum, Al, and lead, Pb, for the creation of electrodes on the superconductive material fabricated from three different oxides Cu_2O, BaO and Y_2O_3 (YBCO). Let us suppose that the enthalpy for formation of one mole of these oxides is

$$\Delta H^0(Y_2O_3) = -455 \text{ kcal/mole}$$
$$\Delta H^0(BaO) = -132 \text{ kcal/mole}$$
$$\Delta H^0(Cu_2O) = -40.8 \text{ kcal/mole}$$

We must choose a metal which creates a stable electrode coating on the YBCO.

(We are sure that you remember that one mole of some material contains an Avogadro number of its atoms, $N_A = 6.02 \cdot 10^{23}$ at/mole and weights in grams the value of which is equal to the atomic number.)

Evidently, we can find from the Ellingham diagram that temperature of the oxidation process for given conditions for Al oxidation will be 1050°C that provides the lead oxidation process:

$$(4/3)Al + O_2 \rightarrow (2/3)Al_2O_3 \qquad \Delta H^0(Al_2O_3) = -200 \text{ kcal/mole}$$
$$2Pb + O_2 \rightarrow 2PbO \qquad \Delta H^0(PbO) = -40.8 \text{ kcal/mole}$$

Reaction of assumed oxides formation are given using the Ellingham diagram as follows:

$$(4/3)Y + O_2 \rightarrow (2/3)Y_2O_3 \qquad \Delta H^0(Y_2O_3) = -455 \text{ kcal/mole}$$
$$2Ba + O_2 \rightarrow 2BaO \qquad \Delta H^0(BaO) = -132 \text{ kcal/mole}$$
$$4Cu + O_2 \rightarrow 2Cu_2O \qquad \Delta H^0(Cu_2O) = -40.8 \text{ kcal/mole}$$

Now we can exclude the oxygen from reactions and estimate the resulting heat effect of reactions for both metals, Al and Pb:

$$(4/3)Al + (2/3)Y_2O_3 \rightarrow (2/3)Al_2O_3 + (4/3)Y \quad \Delta H^0 = -255 \text{ kcal/mole}$$
$$(4/3)Al + (2/3)Y_2O_3 \rightarrow (2/3)Al_2O_3 + (4/3)Y \quad \Delta H^0 = +68 \text{ kcal/mole}$$
$$(4/3)Al + (2/3)Y_2O_3 \rightarrow (2/3)Al_2O_3 + (4/3)Y \quad \Delta H^0 = +159.2 \text{ kcal/mole}$$
$$2Pb + (2/3)Y_2O_3 \rightarrow 2PbO + (4/3)Y \quad \Delta H^0 = -404.2 \text{ kcal/mole}$$

$$2Pb + 2BaO \rightarrow 2PbO + 2Ba \qquad \Delta H^0 = -81.2 \text{ kcal/mole}$$

$$2Pb + 2Cu_2O \rightarrow 2PbO + 4Cu \qquad \Delta H^0 = 0 \text{ kcal/mole}$$

Comparing of results shows that probability of spontaneous reactions between lead and components of YBCO is more than that with aluminum. So, lead can redux the metals from the substrate and oxidize. Therefore, aluminum should be chosen as the electrode layer for YBCO.

1.4 Substrate preparation and cleaning

Now, we will consider the substrate preparation for thin films deposition. First of all, the substrate should have the suitable dimensions for using in the experiments. Usually, we use silicon and glass or quartz substrates of the rectangular form with the dimensions 25×25 mm^2. A diamond cutter is used to cut the substrates. If the quartz and glass pieces are enough lasting materials, the silicon wafers are very brittle. Tweezers should be used to handle the silicon wafers. To cut the silicon wafer, a diamond cutter should be placed on the very edge of the silicon wafer, pressed down firmly, to draw a line with the cutter in parallel to the crystalline direction. The Si wafer should cleave easily. The Si wafer can cleave either as triangle or squares depending on the wafer orientation in the Si crystal lattice. For example, the (100) plane is shown in Fig. 1.10, representing the non-polished back side of the Si wafer decorated by thin gold coating. This picture was observed using an optical microscope with magnification of ×2400.

Following pictures (see Fig. 1.11) illustrate the cutting process. Using a diamond scriber or a sharp tweezers and two straightedges, it is possible to break a wafer precisely along a crystographic plane.

The crystalline direction may be found on the silicon wafers according to the specific flats, as shown in Fig. 1.12. These flats define the crystalline orientation and the type of the doping. Unfortunately, not all wafers are prepared with orientation flats.

Prepared substrates should be cleaned. The cleanliness of the substrate surface exerts a decisive influence on the film growth and adhesion. The process of substrate cleaning requires that bonds are broken

Fig. 1.10 The back side of a Si wafer decorated by Au, oriented with the plane (100).

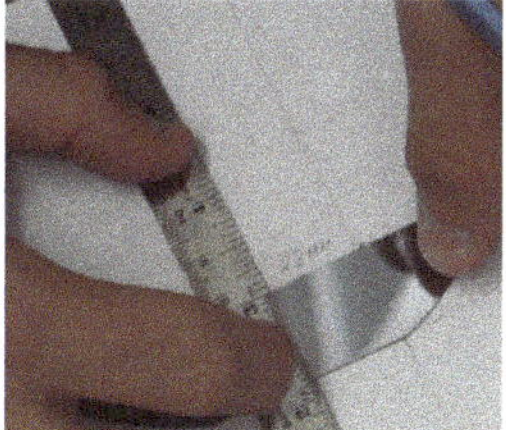

Fig. 1.11 Si wafer cutting on the defined pieces.

between contaminant molecules as well as between the contaminant and the substrate. This may be accomplished by chemical means as in solvent cleaning or by supplying sufficient energy to vaporize the impurity, for example, by heating or particle bombardment. So, all cleaning methods may be divided into three groups:

1. Solvent cleaning using liquid chemistry.
2. Substrate cleaning by heating in high vacuum or in an inert gas environment.
3. Glow discharge cleaning at low-pressure conditions in an inert gas environment.

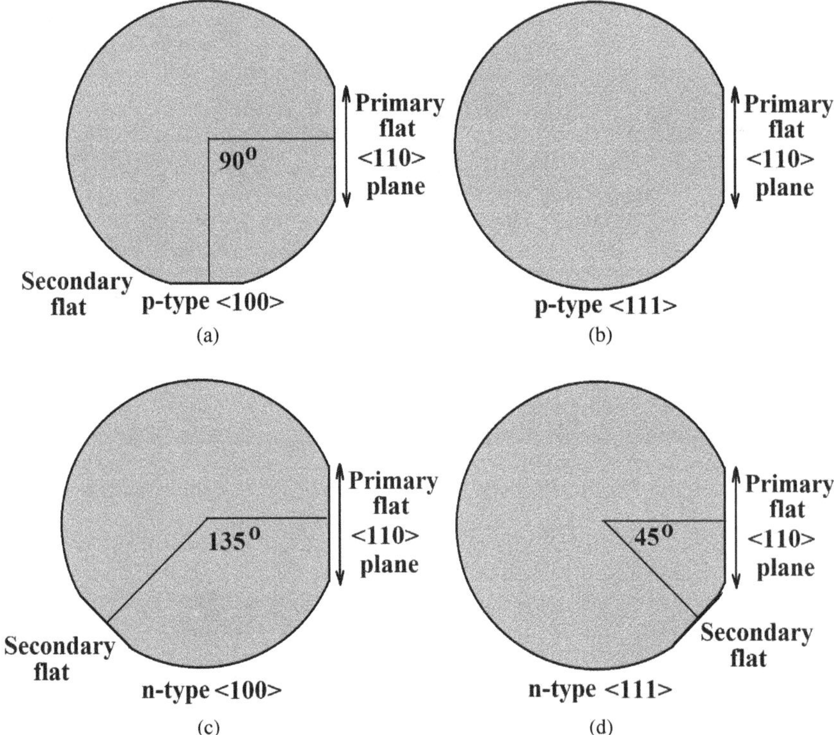

Fig. 1.12 Si wafer orientation and its type definition.

The chemical solvent cleaning enables us to remove organic, oxide and metallic contaminants from the substrate surfaces. Unfortunately, after chemical cleaning, the surfaces have some residue of chemical components adsorbed by them. In addition, a clean surface immediately adsorbs various gas molecules from the atmosphere, which can only be permanently removed by heating in vacuum. Thus, only adsorbed hydrogen molecules remain on the substrate purified by heating. They can only be removed with a glow discharge. Substrate heating in vacuum allows removal of all gas molecules adsorbed from the atmosphere, except the hydrogen molecules. Glow-discharge cleaning allows removing all types of molecules and what remains is only the pure surface.

There are various solvent cleaning methods for different substrates. Solvents can clean oils and organic residues that appear on glass surfaces. For this, acetone or something similar is usually applied

Fig. 1.13 A developing tray and an ultrasound bath for solvent cleaning processes.

together with mechanical scrubbing. This process may be realized using an ultrasound bath and a developing tray as shown in Fig. 1.13.

Unfortunately, solvents themselves (especially acetone) leave their own residues. This is why a two-solvent method is used. This method may be realized by the following way:

- Pour the acetone into a glass container.
- Pour the methanol or isopropanol in a separate container.
- Place the acetone on a hot plate to warm up (do not exceed 55°C).
- Place the silicon or glass substrate in the warm acetone bath for 10 min.
- Remove and place in methanol or isopropanol for 2–5 min.
- Remove and rinse in deionized water (DI water).
- Blow dry with nitrogen or pure compressed air.

If the solvents are clean after the process and you intend to use them again, store in an appropriately labeled container. If not, pour the used acetone and methanol in the solvent waste container. For cleaning the silicon or other semiconductor surfaces, a three-stage process called RCA cleaning is used. These process removes all types of contaminants:

1. Organic cleaning: removal of insoluble organic contaminants with the solution of 5:1:1 = H_2O:H_2O_2:NH_4OH.

2. Oxide layers: Removal of a thin silicon dioxide layer where metallic contaminants may have accumulated as a result of the first stage, using a diluted 20:1 = H_2O:HF solution.
3. Ionic Clean: Removal of ionic and heavy metal atomic contaminants using a solution of 6:1:1 = H_2O:H_2O_2:HCl.

Chapter 2

Gas laws and vacuum technique

2.1 Introduction

There are various methods to prepare thin films. Thin films may be obtained by cutting of the bulk material or by condensation of atoms or molecules from gas or liquid state. However, all cutting methods do not enable production of thin films thinner than ~150–200 μm. Therefore, we should use only such methods which provide for atoms and molecules the possibility to get close to one another. These methods may be realized only in gas or liquid conditions (wet chemistry). In our work, we will consider methods of thin films preparation in gas conditions only.

All gas molecules move in the free state in all directions randomly, Brown's movement. Figure 2.1 illustrates the paths passed by free molecules in the volume.

This movement is limited by collisions between molecules and molecules with walls bounding the gas volume. An average distance between collisions is called the mean free path, λ, and it is one of significant parameters characterizing the gas conditions. Due to collisions and momentum transfer by these molecules, a gas pressure is produced. This pressure is defined by a concentration of molecules and their average energy:

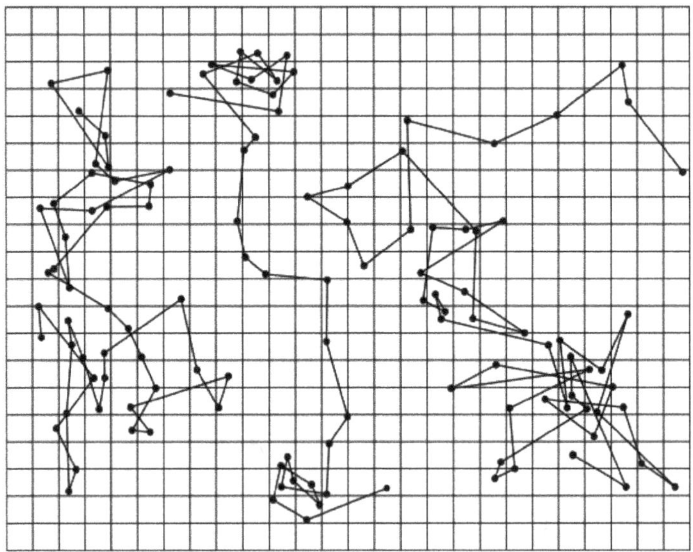

Fig. 2.1 Paths passed by molecules in the volume due to collisions with other molecules.

$$p = \frac{1}{3}\frac{nM}{N_A}\left\langle v^2 \right\rangle = \frac{nRT}{N_A} = \frac{N}{N_A}\frac{RT}{V} = N\frac{k_B T}{V} \qquad (2.1)$$

where n is a concentration of gas molecule, M is a molecular mass of molecules in g/mole, v is a molecular velocity, $N_A = 6.02 \cdot 10^{23}$ mole^{-1} is the Avogadro number, k_B is the Boltzmann constant, $R = N_A \cdot k_B =$ 8.31 J/mole · K is the universal gas constant, N is a number of moles of the gas and T is the absolute environment temperature in K. Evidently, we remember that one mole of a gas is an amount of the gas containing a number of molecules equal to the Avogadro number. Moreover, the weight of one mole of any substance is equal to its atomic or molecular weight in grams. For example, the weight of one mole of carbon C^{12} is equal to 12 g/mole. Thus, we can define a gas pressure using the number of molecules in the volume unit and environment temperature. The gas pressure in the limited volume lower than the atmospheric pressure may be called a **vacuum**. There are various units for pressure measurement, however in the vacuum technique, usually Pa and Torr are applied: **1 atm = 760 mmHg = 760 Torr = 1.013 · 10⁵ Pa**. The conditions at atmospheric pressure and room temperature are called the **normal conditions**. One mole of any gas in the normal conditions takes a volume of V_0 = 22.4 liter. Accordingly, a concentration molecule of air in the unit of volume at normal conditions will be equal to Loschmidt's number N_L: $N_L = \frac{N_A}{V_0} = 2.7 \cdot 10^{19}\,\text{cm}^{-3}$.

Atmospheric air contains two basic components: ~21% of oxygen and ~78% of nitrogen. All residue gases are argon, carbon dioxide and many other gases in small amounts, however influencing the properties of materials and the chemical reactions with them. For example, one mole of air contains ~0.00017% of methane (CH_4) or $0.00017 \cdot 6.02 \cdot 10^{23} = 4.2 \cdot 10^{18}$ mole^{-1}, that represents a significant amount of the reactive material. Therefore, to create pure or stoichiometric thin films without various impurities, it is required to decrease their amount in the reaction volume. That may be done by using the vacuum condition. This is the main reason for application of the vacuum chambers equipped by pumping devices and suitable gas pressure measurement instruments. So, to deposit a thin film on the cleaned surface of a substrate, we should pump the reaction volume up to residual pressure and by requirement insert in this volume a required gas which may also be reactive. Creation of and using the vacuum reaction system requires an understanding of the gas properties, processes occurring in the gas and principles of the moving of the gas.

2.2 Physical properties of gases

As known, behavior of each gas obeys several empirical basic gas laws:

- **Boyle's law:** pV = const, iso-thermal behavior (at constant temperature);
- **Charles's law:** $p = p_0(1+\alpha T)$, iso-choric behavior (at constant volume);
- **Gay-Lussac's law:** $V = V_0\,(1 + \alpha T)$, iso-barial behavior (at constant pressure);
- **Avogadro's law:** $\rho_1/\rho_2 = M_1/M_2$, the gases' density is proportional to their masses;

- **Dalton's law:** every gas is a vacuum to every other gas;
- **Graham's law:** $r_1/r_2 = \sqrt{(M_1/M_2)}$, evaporation rate is proportional to the root square of the gas mass;
- **Ideal gas equation:** $pV = RT, p_\Sigma = \frac{RT}{V}\Sigma_i p_i$, total pressure of various gases in the same volume is a sum of partial pressures;
- **Van der Waals equation**: $(p + a/V^2)(V - b) = RT$.

Here, α is the thermal expansion coefficient, a/V^2 is a constant whose value depends on the gas and characterizes the effective collision area, and b is the volume that is occupied by one mole of the gas molecules.

All molecules move with different velocities under the free Brown movement. The distribution of molecular velocities may be precisely enough described using the Maxwell-Boltzmann distribution:

$$f(v) = \frac{1}{n}\frac{dn}{dv} = \frac{4}{\sqrt{\pi}}\left(\frac{M}{2RT}\right)^{1.5} v^2 e^{-\frac{Mv^2}{2RT}} \quad (2.2)$$

Here, $f(v)dv$ is the probability to find an arbitrary molecule with a velocity inside the interval $(v, v + dv)$. This function, $f(v)$, depends on two important parameters of a gas, its molecular weight and environment temperature. Figure 2.2 illustrates these dependencies.

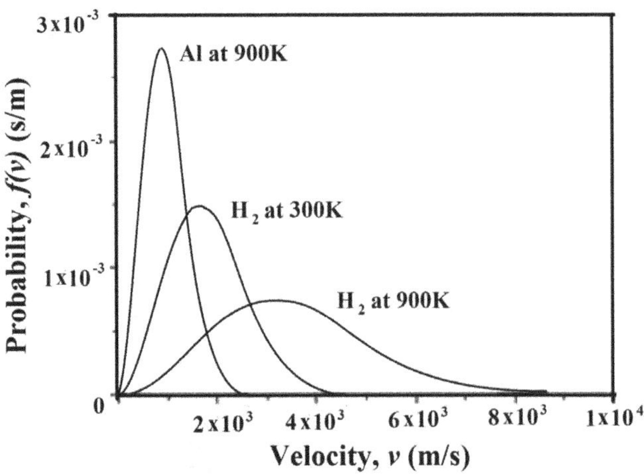

Fig. 2.2 Velocity distribution for aluminum vapor and hydrogen gas.

As shown in Fig. 2.2, increase of the molecular mass leads to increasing the number of high-rate molecules. At the same time, increase of the temperature averages the velocity distribution of the same molecules. Evidently, an integral of the probability function is equal to 1 in all velocity variation intervals:

$$\int_0^\infty f(v)\,dv = 1 \tag{2.3}$$

Therefore, we can define several types of velocities using relation (2.3). These various types of velocities are used sometimes:

- Most probable velocity,

$$v_m : \frac{df(v)}{dv} = 0 \rightarrow v_m = \sqrt{\frac{2RT}{M}} \tag{2.4}$$

- Average velocity,

$$\bar{v} : \bar{v} = \int_0^\infty v f(v)\,dv \Big/ \int_0^\infty f(v)\,dv = \sqrt{\frac{8RT}{\pi M}} \tag{2.5}$$

- Root mean square (rms) velocity:

$$\bar{v}^2 = \int_0^\infty v^2 f(v)\,dv \Big/ \int_0^\infty f(v)\,dv = \frac{3RT}{M} \tag{2.6}$$

All these velocities are related as follows:

$$v_m < \bar{v} < \sqrt{\bar{v}^2} \tag{2.7}$$

As shown, a molecular velocity is defined by atomic mass and environment temperature.

The mean free part between collisions is a characterizing parameter of the gas pressure. This parameter depends on the atomic dimensions and the number of atoms or molecules in the volume. On the other hand, the number of atoms in gas defines the gas pressure. Thus, we can calculate the mean free path of atoms or molecules in some volume as a function of the gas pressure. Two molecules can

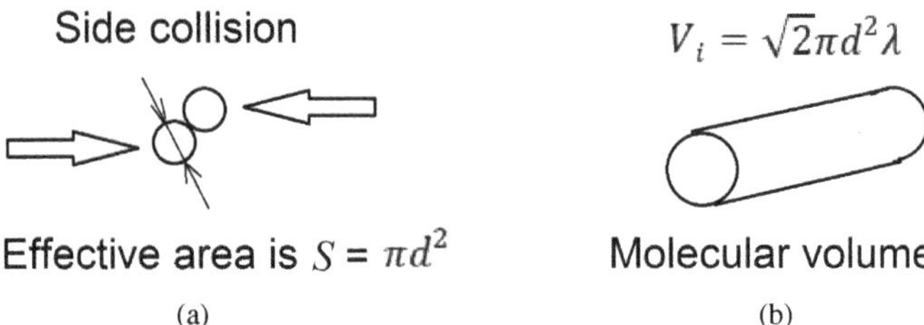

Fig. 2.3 Illustrations for the definitions of effective collision area (a) and molecular volume (b).

collide in several ways: this may be a head-on collision or a side collision. So, one can define an effective area of collision as an area equal to a disc with a double radius of the molecule, $S = \pi d^2$, when d is the molecule's diameter. Figure 2.3 illustrates our calculation.

Here, an effective collision section takes an area with a radius equal to the diameter of the gas molecule as shown in Fig. 2.3(a). A calculated volume of one molecule, V_i, represents a cylinder with the diameter equal to the effective collision area and the length defined by the average distance traveled by the molecule between two collisions. During the same time, each molecule goes the same average distance λ, on a mean free path, if their velocity is equal to an average velocity. Thus, a mean free path may be defined as follows:

$$\lambda = \overline{v}t, \text{ when } \overline{v} \text{ is an average velocity} \tag{2.8}$$

Each molecule moves with a velocity relative to another molecule. Thus, a relative path traveled by molecules will be equal to $\overline{v_{\text{rel}}t} = \sqrt{2}\overline{v}t$ and a volume taken up each time by one molecule will be equal to

$$V_i = \pi d^2 \overline{v_{\text{rel}}t} = \sqrt{2}\pi\, d^2 \overline{v}t = \sqrt{2}\pi\, d^2 \lambda \tag{2.9}$$

The full gas volume represents a sum of volumes of each molecule in this volume at the same time. Therefore, taking into account the relation (2.1), we obtain

$$V = n\sqrt{2}\pi\, d^2 \lambda = \frac{nRT}{pN_A} \text{ or } \lambda = \frac{RT}{p\sqrt{2}\pi\, d^2 N_A} \tag{2.10}$$

As shown, a mean free path is defined by the type of the gas molecule (diameter), by the environment temperature and by the number of molecules in the volume (pressure). Usually, at room temperature (RT = 300 K), a simplified approximation formula for estimation of the mean free path as a gas pressure function may be applied:

$$\lambda = \frac{5 \cdot 10^{-3}}{p}, \text{where } p = [\text{Torr}] \text{ and } \lambda = [\text{cm}] \qquad (2.11)$$

Table 2.1 illustrates a relation between pressure and mean free path at room temperature.

Table 2.1 Relation between pressure and mean free path at room temperature.

Pressure (p)	Mean free path (λ)
1 atm = 760 Torr	66 nm
1 Torr	50 μm
1×10^{-3} Torr	5 cm
1×10^{-6} Torr	50 m
1×10^{-9} Torr	50 km

Another significant parameter used for characterization of the pressure state in the vacuum chamber and defining thin film growth conditions is the molecular impingement flux and related with it the monolayer formation time. This flux, Φ, measured in [$cm^{-2}s^{-1}$], represents a quantity of molecules that can reach the surface unit in the same time or a number of molecules moving in the same direction. Thus, an impingement flux or impingement rate will indicate a number of collisions of moving molecules with average velocity with the surface area:

$$\Phi = \frac{n\bar{v}}{4} \qquad (2.12)$$

After substitution of the average velocity from (2.5) and the number of molecules from (2.1), we obtain:

$$\Phi = \frac{n\bar{v}}{4} = \frac{pN_A}{4RT}\sqrt{\frac{8RT}{\pi M}} = \frac{pN_A}{\sqrt{2\pi RTM}} \approx 3.53 \cdot 10^{22} \frac{p}{\sqrt{MT}}\left[\frac{\text{mol}}{\text{cm}^2\text{s}}\right] \qquad (2.13)$$

An average coverage time, representing a time required for a monolayer formation, is an additional parameter, related with the impingement flux. This time may be calculated as a value inversely proportional to the impingement rate. If *a* designates a number of spaces taken by one molecule per surface area unit, a monolayer formation time, τ, will be calculated as follows:

$$\tau = \frac{a}{\Phi} = \frac{a\sqrt{2\pi RTM}}{pN_A} \approx 4 \cdot 10^{-8}\frac{\sqrt{MT}}{p} \tag{2.14}$$

where *a*, for example, designates the space with diameter of 3Å.

In general, all described parameters may be presented on one diagram enabling calculating all of them on the base of a taken gas pressure in the vacuum chamber or in the space (see Fig. 2.4). This diagram represents a nomogram for graphical calculation without using the theoretical equations. An input parameter, a pressure, for calculation here is presented in two basic units: Torr and Pa at room temperature. All other parameters may be found from the nomogram using simple straightedge, as shown in the diagram. It should be noted that multiplication of a pressure in the vacuum chamber

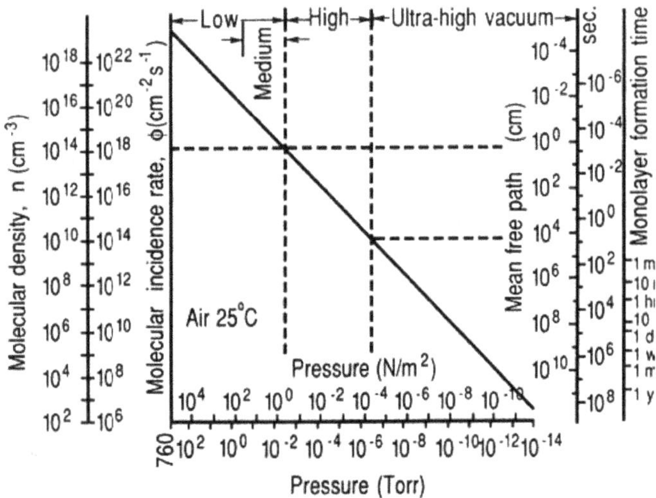

Fig. 2.4. Nomogram for graphical calculation basic gas parameters.

on the monolayer formation time is called an exposure to a surface, which can be measured in the unit called Langmuir. It is named after the American physicist Irving Langmuir: 10^{-6} Torr × 1 s = 1 L (Langmuir).

2.3 Vacuum and gas movement

Gas will always flow where there is a difference in pressure. Assuming that a gas with concentration *n* is in the limited volume with the pressure more than in the environment. This volume has an orifice with area of *A* as shown in Fig. 2.5. Then, these molecules begin to run away from the defined volume with the volume velocity described by Knudsen's equation

$$v = \frac{\varnothing A}{n}\left[\frac{\text{cm}^3}{\text{s}}\right] \quad (2.15)$$

where ∅ is a molecular flux (see Eq. (2.13)).

A gas movement may be organized in the limited volume, for example, in the chamber or in the room. For this, a room should be equipped with suitable pumps and fans. The gas flow depends on the volume dimensions and the pressure inside the volume. As we know, a mean free path characterizes a pressure inside the volume. A relation between volume or system dimensions and mean free path may be described using a coefficient called Knudsen's number:

$$K_n = \frac{D}{\lambda} \quad (2.16)$$

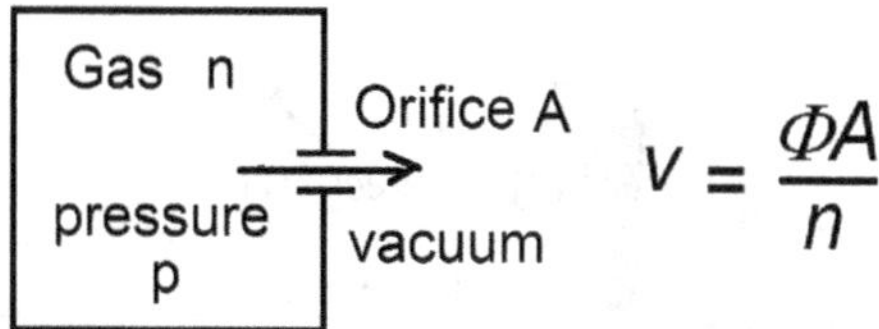

Fig. 2.5 Illustration of the process of gas leaving from the chamber.

where D is the characteristic dimension. The Knudsen number defines a type of gas flow:

1. If $\lambda \gg D$ ($K_n < 1$), the gas movement is called a molecular or laminar flow. This flow may be realized, for example, in the vacuum chamber at very low pressure, when collisions with container walls prevail. In this case, gas does not mix when moving.
2. $1 < K_n < 100$, this movement is called an intermediate flow.
3. $\lambda \ll D$ ($K_n > 100$), the gas movement is called a viscous or turbulence flow. In this case, the molecule-to-molecule collisions prevail. Gas mix when moving, which create good conditions for gas–chemical reactions.

When we want to use the vacuum condition for some technology, the specific requirements for the gas movement and pumping techniques appear. These requirements are a short pumping time, low cost of the pumps and connecting systems, net directed movement of a gas, low back-streaming, limitation on the back-streaming gas composition, etc. Figure 2.6 represents a schematic construction of a typical high-vacuum system, equipped with a two-stage vacuum pumping arrangement.

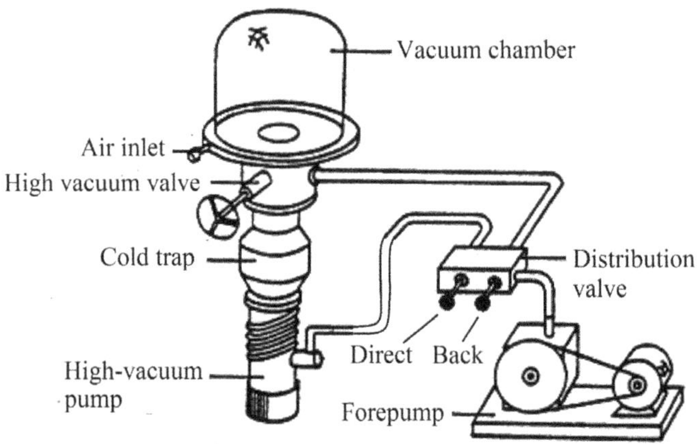

Fig. 2.6 Schematic construction of a typical high-vacuum system.

The vacuum system presented in Fig. 2.6 consists of a vacuum chamber, a high-vacuum pump, a low-vacuum pump (forepump), a high-vacuum valve, a distribution valve and connecting pipes. A main goal for creation and application of a high-vacuum system is formation of the specific pure conditions in the vacuum chamber. We need to remove air or other gases from the vacuum chamber, to keep the high-vacuum conditions inside the chamber or to fill it with the required gas.

Each vacuum pump works only in the limited diapason of pressures. Therefore, we can approximately define these work pressure areas (see Fig. 2.4):

1. A low, rough or fore vacuum, where pressure is in the range of 760–10^{-3} Torr;
2. A high-vacuum area, where pressure is in the interval 10^{-3}–10^{-6} Torr;
3. A very low pressure or very high vacuum, where pressure is lower than 10^{-6} Torr.

Consequently, for each vacuum diapason, there are suitable vacuum pumps and vacuum measurement gauges, which we'll consider later.

A pumping speed is defined by the type of vacuum pump, the geometry of the vacuum chamber, the configuration of the vacuum system and by the geometry of connecting pipes. The connecting pipes system may be characterized by a flow conductance, *C*, measured in the volume of a gas, *V*, going through a system: m^3/s or l/s. This parameter, a flow conductance, may be measured as a relation of a gas throughput, Q = [pressure × volume/s], to the pressure difference between two ends of a pipe, Δp:

$$C = \frac{Q}{\Delta p}\left[\frac{m^3}{s}\right] \tag{2.17}$$

Each element of the vacuum connecting system has own flow conductivity due to different geometry of the elements, C_i. The pressure fall occurs in each element of the system. So, the total flow conductance of the vacuum system will be equal to the sum of each conductive element. Like the electrical circuits, we can designate a resistance of vacuum pipes or vacuum system elements as a reciprocal value to the conductance value: $R = 1/C$. Now, Eq. (2.17) may be rewritten as follows:

$$Q = \frac{\Delta p}{R} \tag{2.18}$$

which provides for in series and in parallel connections the following relations:

$$\begin{cases} R_\Sigma = R_1 + R_2 + R_3 + \dots \text{ in series} \\ \dfrac{1}{R_\Sigma} = \dfrac{1}{R_1} + \dfrac{1}{R_1} + \dfrac{1}{R_1} + \dots \text{ in parallel} \end{cases} \tag{2.19}$$

A flow conductance depends on geometry and pressure difference between inlet and outlet of an element. However, in the case of laminar flow, a flow conductance will be independent of the pressure. Thus, the flow conductance of an orifice with area of A [cm^2] for air at room temperature and in the case of laminar flow will be equal to $C = 11.7A$. Figure 2.7 presents the conductance for elements with various simple geometries. Evidently, according to the relations (2.19), we can write analogous equations for the conductance:

$$\begin{cases} C_\Sigma = \sum_i C_i \text{ for conductances joined in parallel} \\ \dfrac{1}{C_\Sigma} = \sum_i \dfrac{1}{C_i} \text{ for conductances joined in series} \end{cases} \tag{2.20}$$

So, to calculate a conductance of the vacuum system, we need to calculate the conductance of each element of the system and using relation (2.20), calculate the resulting conductance. Calculation and experience say that there is a rule for designing the vacuum systems: **the lines should be as short and as wide as possible** to provide a viable and satisfactory vacuum system.

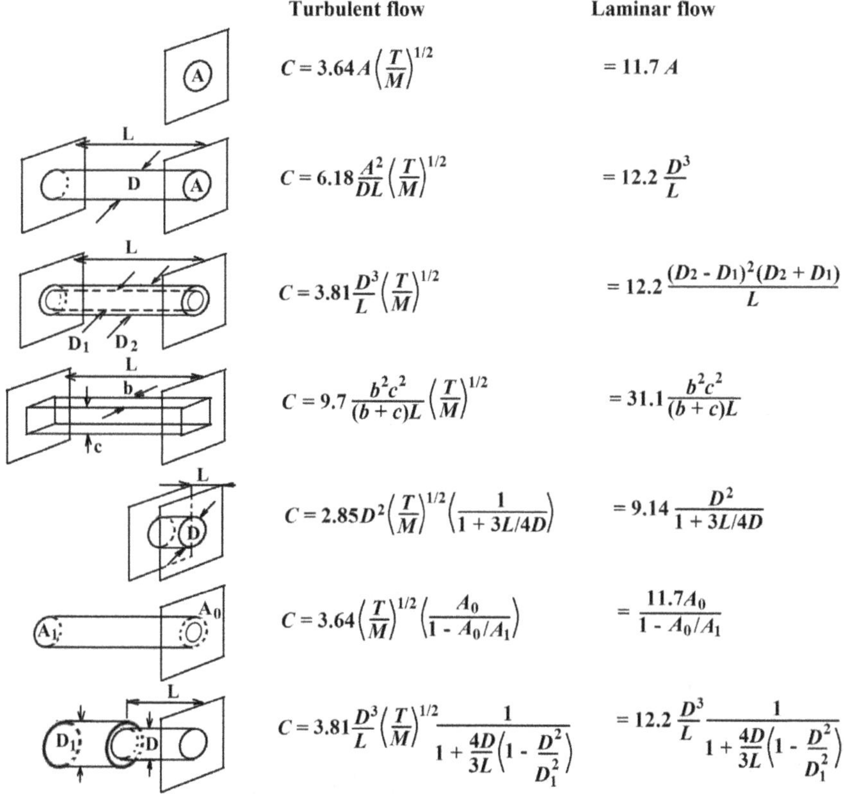

Fig. 2.7 Flow conductance for various simple elements of a vacuum system.

A pumping speed, S, measured on the output of the vessel, is an important parameter of the vacuum system. It is defined by the relation of a gas throughput Q to the pressure at the pump inlet, p, or the gas velocity v through the area A of the cross-section at the mouth of the pump or the volumetric amount of gas V evacuated from the chamber in the time unit:

$$S = \frac{Q}{p} = Av = \frac{dV}{dt}\left[\frac{\mathrm{m}^3}{\mathrm{s}}\right] \tag{2.21}$$

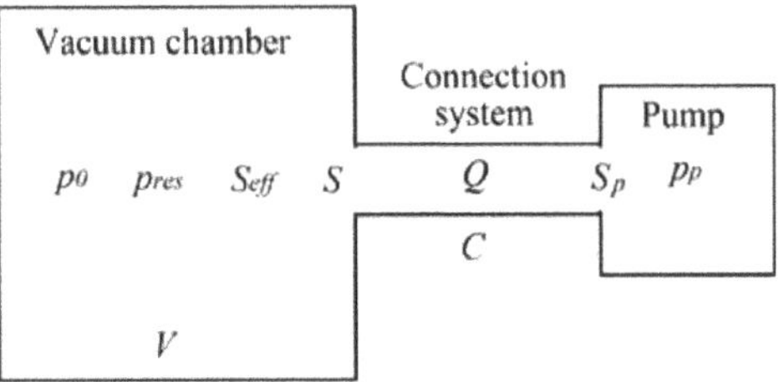

Fig. 2.8 Typical simple vacuum system.

From this equation and substituting relation (2.1), we obtain the equation describing the throughput in terms of molecules per unit volume:

$$Q = \frac{k_B T dN}{dt} = \frac{d(pV)}{dt} = V\frac{dp}{dt}\left[\mathrm{Pa}\cdot\mathrm{m}^3\cdot\mathrm{s}^{-1} = \mathrm{N}\cdot\mathrm{m}\cdot\mathrm{s}^{-1} = \mathrm{J}\cdot\mathrm{s}^{-1} = \mathrm{W}\right] \quad (2.22)$$

Let us consider a simple vacuum system as shown in Fig. 2.8. Here, a vacuum chamber with volume V, connected with the pump through a connecting pipe-system, which has a flow conductance C, is evacuated from an initial pressure p_0 up to the residual pressure p_{res}. A pump produces the back-streaming flow with the throughput Q_p. Pressure in the inlet to the pump is p_p. S_p is the pumping speed at the inlet of the pump. The throughput Q of the system is equal to $Q = C(p - p_p) = pS = p_p S_p$, if we assume that $Q_p = 0$. From this relation, we will obtain the following equation for the pumping speed:

$$S = S_p\frac{p_p}{p} = S_p\frac{p_p}{p_p\left(\frac{S_p}{C}+1\right)} = \frac{S_p}{\frac{S_p}{C}+1} \quad (2.23)$$

As shown in Eq. (2.23), the pumping speed will be higher if the flow conductance, C, will be more. In realty the $Q_p \neq 0$, therefore the pumping process will be described by the following equation called the Continuity equation:

$$Q = S_p p - Q_p = S_p p\left(1 - \frac{Q_p}{S_p p}\right) \tag{2.24}$$

Now, the residual pressure p_{res} will be reached when the throughput decreases to zero, $Q \rightarrow 0$. So, $Q_p = S_p p_{res}$ and the effective pumping speed will fall to zero:

$$S = s_{eff} = \frac{Q}{p} = S_p\left(1 - \frac{p_{res}}{p}\right) \tag{2.25}$$

Evidently, the pumping speed decreases while pumping due to fall of the pressure inside the vacuum chamber. Thus, the minimum or ultimate or residual pressure in the chamber will be reached at the moment the pumping speed comes to the minimum and pressure decrease becomes equal to the backflow from the pump and various leakages. The evacuation process may be described also by the continuity equation in the differential form:

$$Vdp = Q_\Sigma dt - Spdt \tag{2.26}$$

where Q_Σ represents the sum of all leaks and backflow throughput, which together define the residual pressure p_{res}. In the ideal case, when $Q_\Sigma = 0$, the solution of Eq. (2.26) shows an exponential decreasing of pressure in the vacuum chamber:

$$p = p_0 e^{-\frac{S}{V}t} \tag{2.27}$$

In real conditions, this solution will be as follows:

$$\frac{p - p_{res}}{p_0 - p_{res}} = e^{-\frac{S}{V}t} \tag{2.28}$$

That enables us to calculate the time required to reach the defined pressure. Figure 2.9 illustrates both solutions of the continuity equation.

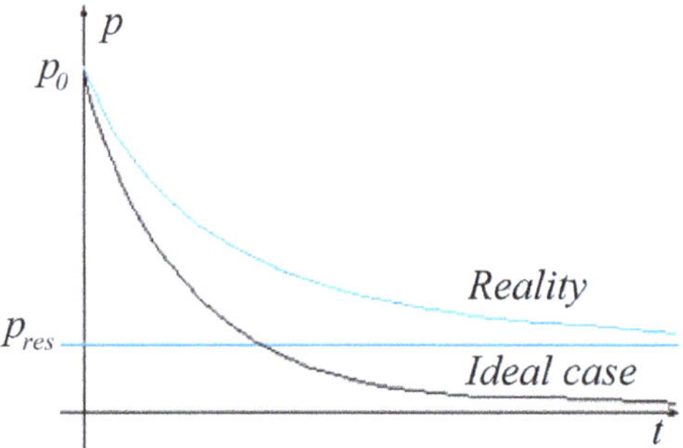

Fig. 2.9 Graphical representation of the continuity equation solutions for ideal and real cases.

To decrease the leakage in the vacuum chamber, only specific materials with low pressure of steams should be applied, for example, a stainless steel is the right material for the vacuum chamber.

2.4 Pressure gauges

A main goal of pumping from vacuum chambers is decreasing the air pressure. Air consists of several gases as shown in Table 2.2. Molecules of different gases have different dimensions and weights, therefore the pumping speed for different gases is different. It is believed that the average molecular weight of air molecules is 29 g/mole. Moreover, the devices that measure the pressure are differently sensitive to different gases. When we want to measure a level of vacuum, we measure the pressure. There are many devices for pressure measurement. Here, we will consider only widespread devices. They are based on different principles and each one is intended for measurements at limited range of pressures only.

Figure 2.10 presents various gauges used for measurement of the pressure in vacuum systems. These measurement devices are based on various principles and used for measurement in different pressure ranges.

For measurement of rough vacuum, we use the vacuum gauges of the following types: Capacitive manometers (Baratron), Thermocouple gauges and Pirani gauges (Convectron).

Table 2.2. Air composition.

Gas	% vol.	% weight	Partial pressure, p [Torr]
N_2	78	75.5	590
O_2	20.95	23.14	160
Ar	0.93	1.25	7
CO_2	0.03	0.05	0.25
Ne	0.002	0.001	0.014
He	$52 \cdot 10^{-4}$	$72 \cdot 10^{-5}$	$4 \cdot 10^{-3}$
H_2	$5 \cdot 10^{-5}$	$5 \cdot 10^{-6}$	$4 \cdot 10^{-4}$

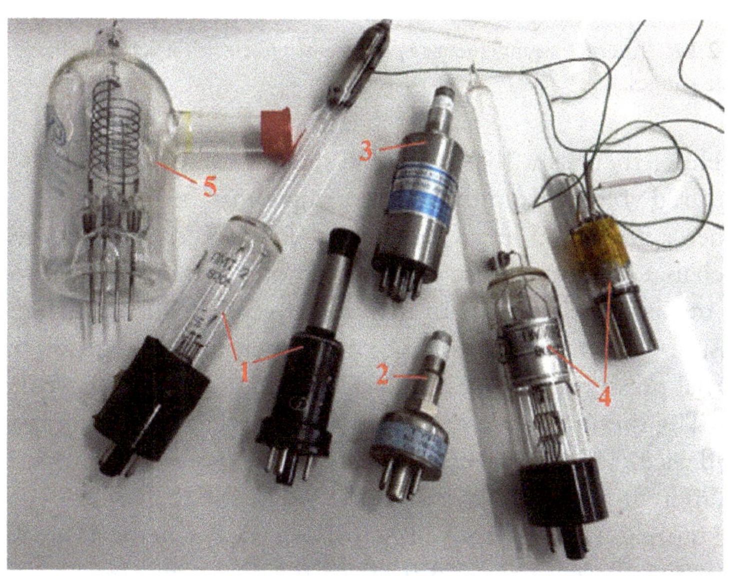

Fig. 2.10 Various sensors used in vacuum systems: (1) Thermocouple gauges, (2) and (3) Pirani gauges, (4) Ion gauges for high vacuum, and (5) Bayert-Alpert gauge.

The thermocouple gauge is the simplest and most widespread device. It is devoted for measurement in the interval from 1 atm up to ~$5 \cdot 10^{-3}$ Torr. Figure 2.11 illustrates the internal arrangement (Fig. 2.11(a)) and the principle electrical scheme (Fig. 2.11(b)) of this device.

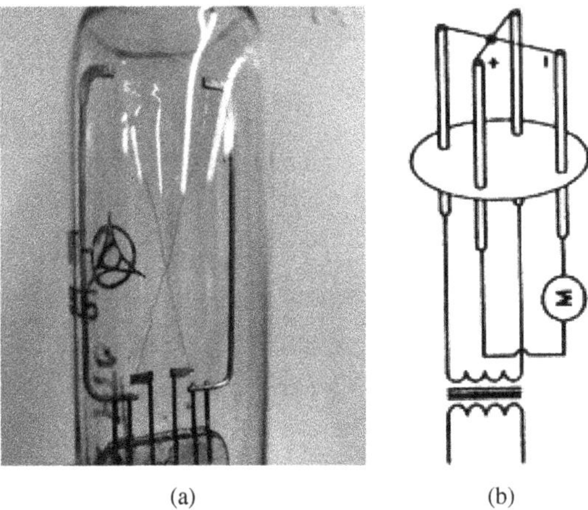

(a) (b)

Fig. 2.11 Internal arrangement (a) and principle electrical scheme (b) of thermocouple gauges.

Here, there is a thermocouple heated using a separated supply. Measured voltage from the thermocouple may be read using a microammeter graduated in pressure units, Pa or Torr. This gauge senses the change in temperature of the common point of the thermocouple due to different amounts of heat removed defined by change in the pressure. The measured voltage of the thermocouple grows with the internal pressure decreases and related decrease of the gas concentration. The thermocouple gauge is a very rugged, reliable and inexpensive device. The disadvantage of this device is its different sensitivity for different gases, however it works well with air.

Figure 2.12 presents a principal electrical scheme (Fig. 2.12(a)) and external view (Fig. 2.12(b)) of the Pirani gauge. As shown in Fig. 2.12(a), this gauge represents an electrical bridge. The measuring resistor arranged in one shoulder of the bridge, situates in the vacuum and heats up to 100–150°C. Before measurements, the bridge is balanced using the compensating resistor. Due to decreasing the pressure, the gas concentration and heat removal decrease and the current flowing through the measuring resistor grows. So, we can read a measured temperature on the ammeter graduated in the pressure units. The Pirani gauge works steadily in the pressure interval from atmospheric pressure up to 10^{-4} Torr. The vacuum gauges of

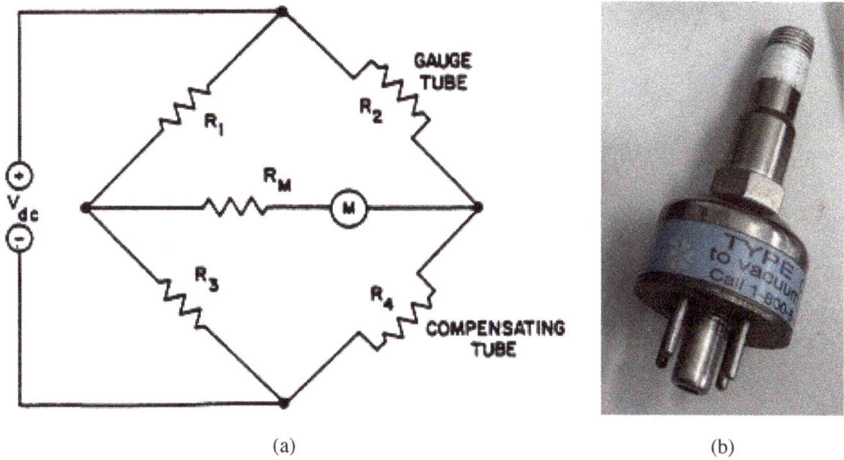

Fig. 2.12 Principle electrical scheme (a) and external view (b) of Pirani gauges.

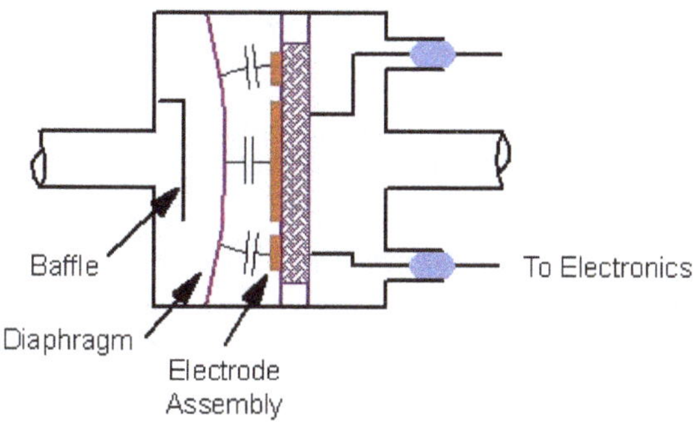

Fig. 2.13 A schematic side view of the capacitance manometer.

both described types are called "Convectron" sometimes due to dependence of measured parameter on the gas convection around the gauge.

A capacitance manometer (Baratron) is intended for the direct measurement of atmospheric pressure down to about 10^{-5} Torr. Figure 2.13 represents a schematic internal side view of the capacitance manometer. The principle of operation of this gauge is based on the comparing of the pressure in the measured volume and in the reference volume pumped up to low pressure. Both these volumes are

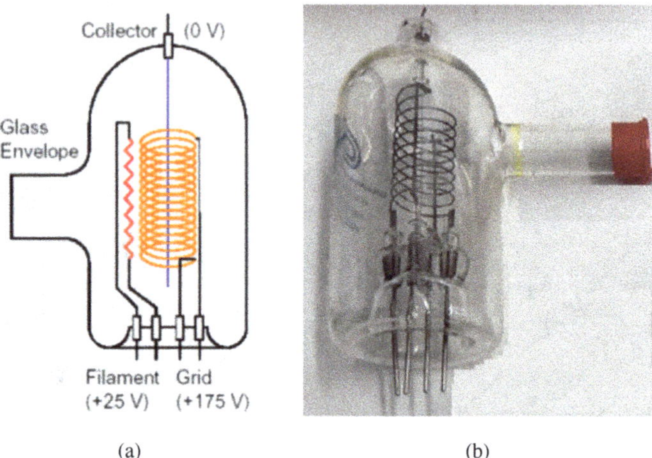

Fig. 2.14 Schematic and real side view of the Bayard-Alpert high-vacuum gauge.

divided by a diaphragm which can move. Difference in the pressure from two sides is converted by electronic module to the readable pressure value.

For the measurement of high-vacuum pressures, ion gauges have been commonly used. Figure 2.14 shows the ion gauge of Bayard-Alpert intended for the measurement of low pressure in the interval from 10^{-2} Torr up to 10^{-9} Torr. This ion gauge is commonly used in vacuum technology. This gauge represents a triode structure. It consists of a heated cathode, such as a tungsten wire emitting electrons, an accelerating grid providing enough energy for ionization of the gas inside the glass bulb connected with the vacuum chamber, and a thin wire biased to collect the ions. An ion current measured in this system is proportional to the gas concentration inside the chamber.

The ion gauge is sensitive to various gases. Figure 2.15 illustrates the relative sensitivity of Bayard-Alpert gauges to different gases.

This property of ion gauges, the different sensitivity to various gases, is used sometimes for definition of leakage places in vacuum systems. An evacuated vacuum system should be sprayed with acetone in suspicious places. The drastic change of the measuring pressure will show the leakage point in this case.

Gas	Relative Sensitivity
H_2	0.42 - 0.53
He	0.18
H_2C	0.9
Ne	0.25
N_2	1.00
CO	1.05 - 1.1
O_2	0.8 - 0.9
Ar	1.2
Hg	3.5
Acetone	5

Fig. 2.15 Approximate relative sensitivity of an ion gauge to different gases.

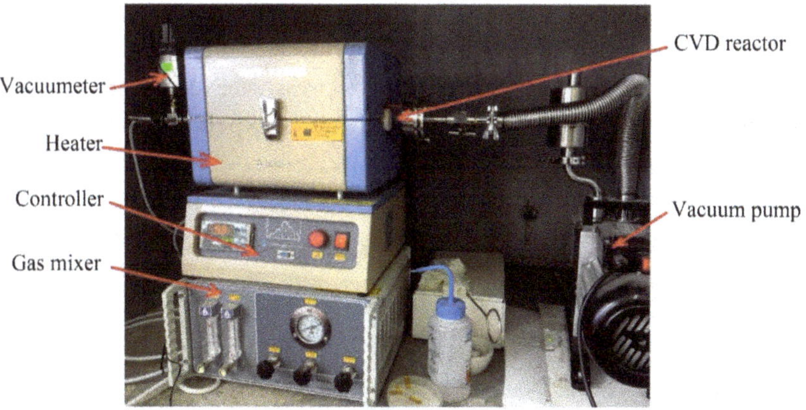

Fig. 2.16 CVD system equipped with the for-vacuum pump.

2.5 Vacuum Pumps

Usually, a vacuum system, evacuating the reactor chamber, consists of vacuum pumps providing the required level of a pressure. For example, the chemical vapor deposition (CVD) systems, working at the rough vacuum level, require only one-stage mechanical rotary vane pump as shown in Fig. 2.16.

The rotary vane or rotary oil pumps are devoted to evacuation of the vacuum chambers or reactors up to the for-vacuum pressure, i.e., up to ~5 · 10^{-3} Torr. These pumps are the most commonly used devices for all basic vacuum applications from atmospheric pressure down to 10^{-3} Torr. These pumps are also used as backing pumps for other high-vacuum gas delivering systems including diffusion pumps and turbomolecular pumps, which we will consider later.

A mechanical rotary pump consists of two main parts: an electrical motor and a pump itself. Figure 2.17 illustrates the external view of the mechanical pump (a) and the schematic section of the pump immersed in the oil bath (b). As shown in the schematic section of the pump, gases are removed here by compressing them slightly above atmospheric pressure and then forcing them through a check valve F. Gas enters the inlet port and is trapped between the rotor vanes and the pump body. The rotor is mounted eccentrically on the motor axis, so the spring-loaded vanes scan the stator surface when the rotor rotates. In this way, the rotating rotor compresses the gas and sweeps it through the discharge port. When gas pressure exceeds atmospheric pressure, the exhaust valve opens and gas is expelled. In this pump, oil is used as lubricant and sealant.

The physical vacuum deposition (PVD) systems usually require significantly lower pressure that may be obtained using two-stage

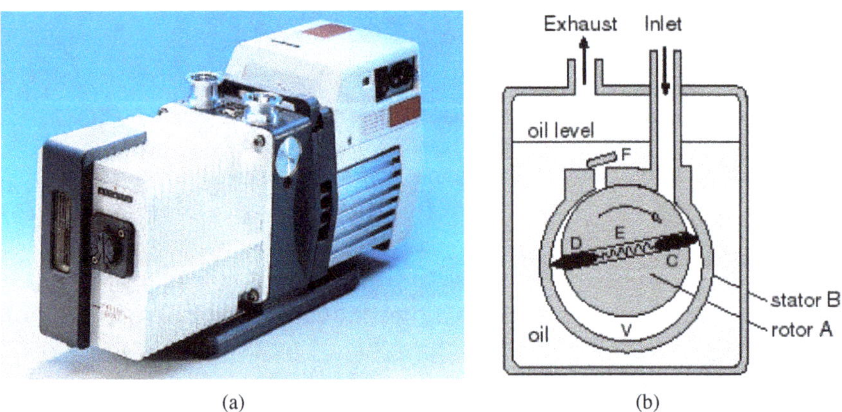

(a) (b)

Fig. 2.17 External view (a) and the schematic section (b) of the mechanical rotary pump.

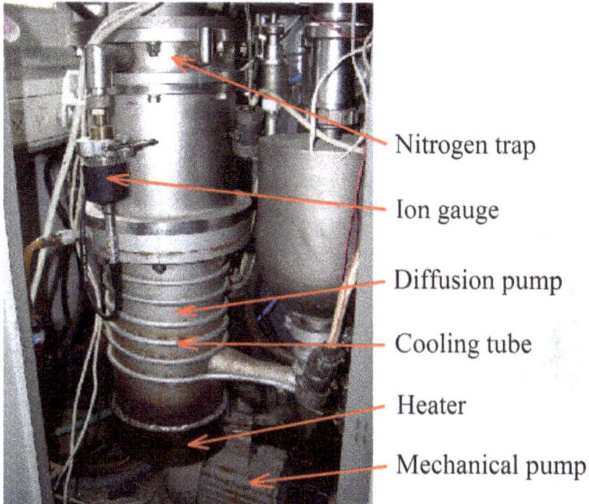

Fig. 2.18 Two-stage vacuum system containing mechanical and diffusion pumps.

vacuum equipment. Such systems enable providing a residual pressure at the level of 10^{-5}–10^{-7} Torr. Figure 2.18 illustrates the two-stage vacuum system.

This system consists of a mechanical rotary vane pump providing the first evacuating stage and a high-vacuum steam-oil (diffusion) pump. Diffusion pump consists of a stainless-steel chamber with water-cooled walls. Inside this chamber, three varying sized, cone-shaped pressure jets are arranged to create the upward directed way for the oil-steams stream. Figure 2.19 illustrates the operation principle of the diffusion pump.

A silicone-based diffusion pump oil (the specific oil with low pressure of steams) is in the bottom of the chamber. After heating up to 180–270°C, the oil boils and converts in to steam which travels with high velocity and captures the air molecules through diffusion. As the walls of the pump chamber are cooled, as the gas reaches the chamber walls, it immediately returns to liquid state releasing the trapped air molecules at a lower position and at increased pressure creating the vacuum. The first-stage mechanical (fore-vacuum) pump evacuates these free air molecules. The oil drips back to the bottom of the pump chamber where it is heated again. The diffusion pump evacuates gas only in the heated state.

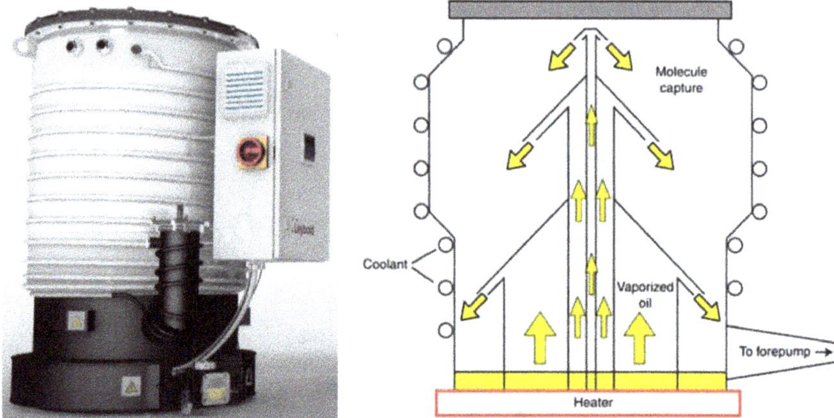

Fig. 2.19 External view and schematic section of the steam-oil (diffusion) pump.

The diffusion pump has relatively high pumping speed for quite low cost and it works without vibrations and noise. However, these pumps are sensitive to mistakes in the operating procedure and require additional time for heating and cooling. To begin pumping with the diffusion pump, a preliminary low pressure (fore vacuum) must be created in the vacuum chamber using the first stage mechanical pump. One of the disadvantages of diffusion pumps is the high back streaming of the oil steam. To protect the evacuating volume from the back streaming, the oil traps are usually placed between the pump and the vacuum chamber. Sometimes, such traps may be specially cooled by water or liquid nitrogen that enable significantly reducing the pressure inside the vacuum chambers and fully protecting from the oil steam.

Another type of two-stage vacuum system is presented in Fig. 2.20. Here, a turbomolecular pump provides high vacuum in the reaction chamber. A mechanical pump is also used in this system to provide the first-stage pumping.

Turbomolecular pumps, also known as "turbo pumps"; their applications cover all processes and vacuum systems in the $10^{-4}/10^{-10}$ Torr pressure range. Turbo pump resembles a jet engine. Several very high-speed rotors, each with multiple blades shaped with angled leading edges, impart momentum to gas molecules in the direction of the

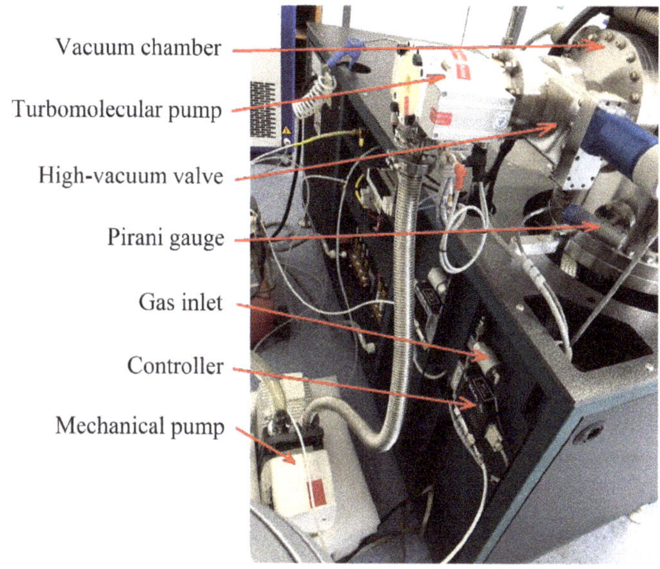

Fig. 2.20 Two-stage vacuum system based on the turbomolecular pump.

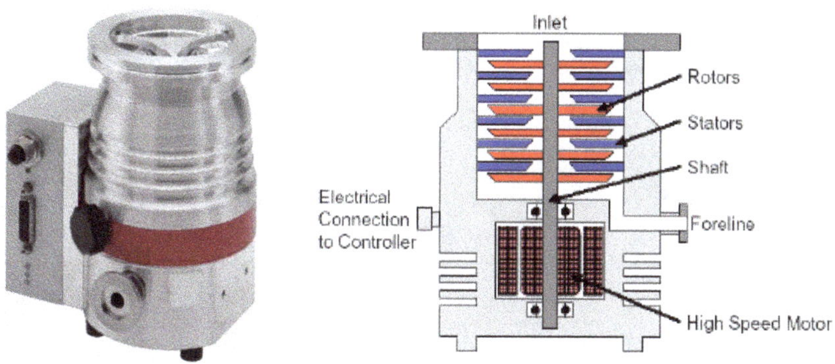

Fig. 2.21 External view of a schematic section of the turbomolecular pump.

next rotor down the stack. The compression ratio across the pump for a mediate molecular weight gas may exceed 10^8. An angular rotation speed of rotors in the turbomolecular pump reaches tens of thousands of rotations per minute. Figure 2.21 illustrates an external view of a schematic section of the turbomolecular pump. It consists of a stainless-steel chamber with stators fixed on its internal wall and rotors arranged on the axis of the high-speed electrical motor. Distance between rotor and nearest stator does not prevail beyond 0.1 mm. Thus, the turbomolecular pump is a precision electromechanical device requiring attentive and proper service by qualified staff. The pumping speed of a two-stage system equipped with the turbomolecular pump is very high. For example, the system presented in Fig. 2.20 can reach the pressure of $2 \cdot 10^{-5}$ Torr in less than an hour.

Turbo pumps have advantages over diffusion pumps. Correctly operated they do not back-stream oil into the vacuum system at any time and can be started and stopped in a few minutes. The last feature means a turbo pump can be sometimes directly connected to the chamber without a high-vacuum valve. This saves money and improves pumping conductance. But turbo pumps can be noisy and they induce vibration. Turbo pumps are also expensive and the compression ratios for hydrogen and helium are low.

Chapter 3

Physical vacuum deposition, sputtering

3.1 Introduction

Our goal is to grow a thin nano-dimensional structure with predefined properties. To solve this problem, we need to understand why a thin film grows, what conditions influence the growth process and how it goes. Here, the science of thermodynamics has been applied to help in the growth of thin films. Thermodynamics studies various transformations of heat into other forms of energy. It deals with macroscopic bodies consisting of great number of particles and with no single, separated molecules. Therefore, thermodynamics is a part of Statistical Physics. In particular, thermodynamics is applied for studying the chemical reactions and phase conversions between different forms of matter. Thermodynamics is based on the four empirical laws discovered in the course of experiments. Let us consider these laws briefly.

The Zeroth Law of Thermodynamics states that if two systems are in thermodynamic equilibrium with a third system, the two original systems are in thermal equilibrium with each other. Max Planck formulated this Law in 1897. This Law introduces the concepts of equilibrium and absolute temperature. On the basis of this Law, one can say that isolated systems with constant external conditions reach the equilibrium state. If there are two isolated systems, *A* and *B*, in contact with each other, they reach the equilibrium state as well. We

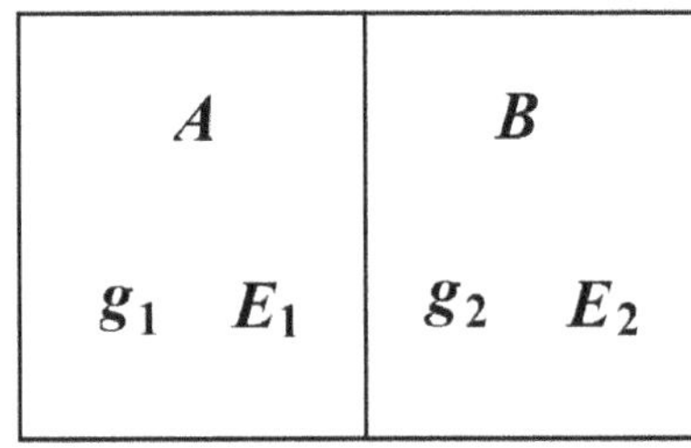

Fig. 3.1 Illustration for the zeroth law of thermodynamics.

designate a number of different states of the systems *A* and *B* as g_1 and g_2, respectively, and E_1 and E_2 are energies of these systems. To specify a concept of the state of the system, we can suggest that this is the number of particles in the system or such is their mutual arrangement. Figure 3.1 illustrates these two systems which are in contact and the internal wall between them may be removed.

When we remove the wall between systems *A* and *B*, the full number of states of the novel common system will be $g = g_1 g_2$ and the full energy of the common system, $E = E_1 + E_2$. As such a number of states is a function of energy, one can write the following relation: $g = g_1(E_1)\, g_2(E - E_1)$. Differentiating this relation gives the following equation:

$$\frac{dg}{dE} = g_2 \frac{dg_1}{dE} - g_1 \frac{dg_2}{dE} \tag{3.1}$$

If our system is in the equilibrium state, so $dg/dE = 0$ and the relation (3.1) transforms:

$$\frac{1}{g_1}\frac{dg_1}{dE} = \frac{1}{g_2}\frac{dg_2}{dE} \quad or \quad \frac{d}{dE}\ln g_1 = \frac{d}{dE}\ln g_2 \tag{3.2}$$

In this equation, $S \equiv \ln g$ is called an <u>**entropy**</u> of the system and a value $T \equiv \frac{1}{\frac{d}{dE}\ln g}$ is called an **absolute temperature** of the system measured in K (Kelvin) degrees. As such an entropy is the logarithmic value, it cannot be lower than zero. Also, an absolute temperature is limited from bottom.

The First Law of Thermodynamics states that energy can be converted from one form to another with the interaction of heat, work and internal energy, but it cannot be created nor destroyed, under any circumstances. This law reflects the generalized energy conservation law. If we designate the full internal energy of the system as U, this law may be described by the following equation:

$$dU - \delta Q = \mp \delta A \tag{3.3}$$

where dU is the change of internal energy of the system, δQ is the elementary quantity of the heat passed to the system and δA is the elementary work done by the system or over the system. The positive sign relates to the work done by the system and the negative sign designates the work done over the system. Sometimes, energy of the mass transfer or a chemical energy may be incorporated into Eq. (3.3). The work in the general equation may be presented by the multiplication of the pressure on the volume. Physical meaning of the concept entropy is in the definition of disorder levels or the non-ideality of the system. The physical meaning of the concept absolute temperature is the possibility to measure the kinetic energy, temperature is the measure of energy.

Look at the following well-known relation:

$$\frac{3}{2}kT = \frac{mv^2}{2} \tag{3.4}$$

Here, we see that kinetic energy is proportional to temperature with Boltzmann's constant as the proportionality coefficient. Accordingly, one can write that 1 eV = 11,600 K = $1.6 \cdot 10^{19}$ J.

The Second Law of Thermodynamics has many definitions, for example: heat cannot transfer from cool body to hot without additional changes in the system; it is impossible to create a perpetual motion machine; the state of entropy of the entire universe, as an isolated system, will always increase over time. The second law also states that the changes in the entropy in the universe can never be negative. In other words, there is a function $S = f(T, N, x)$ for each system called entropy which depends on temperature T, with number of particles in the system N, and position of these particles x. According to the definition of temperature, each change in the temperature leads to a change in entropy:

$$\delta Q = TdS \tag{3.5}$$

The Third Law of Thermodynamics says that both temperature and entropy are limited from the bottom, or if $T = 0$, then $S = 0$ also. This means that a totally perfect (100% pure) crystalline structure, at absolute zero (0 Kelvin), will have no entropy (S). The non-ideal crystalline structures have entropy according to the disorders of the systems.

Let us return to the question, why thin films grow? The films consist of atoms, so the nucleation process should precede the growth. Moreover, before growing the thin films, we separate the matter in the atoms or molecules. Thus, the gas state of matter is the precursor for the thin film growth. In other word, we have the thermodynamic process of phase conversion between the gas state and the solid state or the condensation process. Figure 3.2 illustrates various processes occurring during the nucleation of the novel thin film. Growth of thin film is a non-equilibrium process. To grow a thin film, a flow of particles going to the substrate surface, J, should be more than the reverse flow, J_{vp}. Relation between direct and reverse flows may be described by the parameter ζ called supersaturation:

$$\zeta = \frac{J}{J_{vp}} = \frac{p}{p_{vp}} \tag{3.6}$$

where p is a pressure of the condensing gas and p_{vp} is a pressure of the gas over the substrate. Both flows are described by quantity of particles moving through the defined area in the time unit as shown in Eq. (2.13). They are dependent on the kinetic energy of particles and pressure on flows:

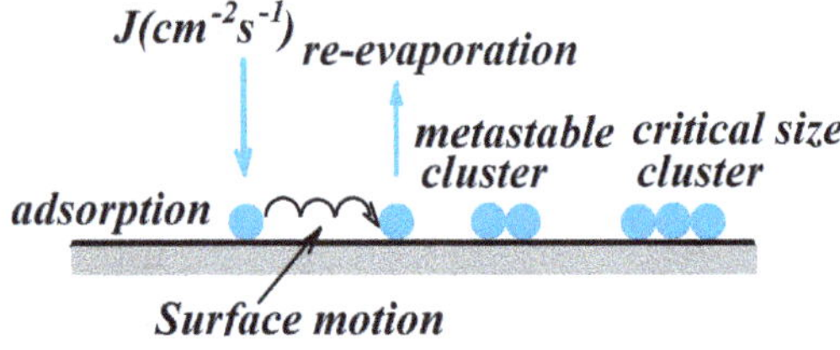

Fig. 3.2 Schematic view of the process occurring during nucleation of a thin film.

$$J = 3.53 \cdot 10^{22} \frac{p}{\sqrt{MT}} \left[\frac{\text{mol}}{\text{cm}^2\text{s}} \right] \quad (3.7)$$

As shown in Fig.3.2, when an atom comes to the substrate, depending on its kinetic energy and temperature of the substrate, it starts to move and search for a place that is energetically beneficial to adsorb. Through this motion, the atom may reevaporate or find another atom or cluster of atoms. Evidently, the group of atoms (cluster) will be stable only if it is more than some critical size. Such a cluster can grow with attachment of additional atoms. The cluster lower than this critical size will decay.

Under condensation process, the Gibbs free energy will decrease by value ΔG_V on the unit volume. Wherein, the free energy of the system will grow due to insertion of new interfaces between the cluster and environment. So, the Gibbs free energy will be the sum of the decreases due to volume changes and increases due to the surface growth:

$$\Delta G = \frac{4}{3}\pi r^3 \Delta G_V + 4\pi r^2 \gamma, \quad (3.8)$$

where γ is an energy of interface on the unit area. Figure 3.3 illustrates the relation (3.8), describing an interaction between two opposite energies.

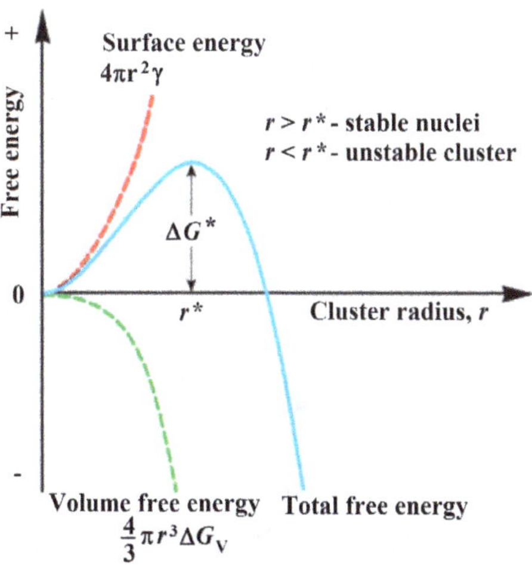

Fig. 3.3 Relation between two opposite free energies.

The critical radius of the cluster may be found from the condition of maximum change in the Gibbs free energy. In these conditions, the cluster forms the spherical shape. Differentiating Eq. (3.8), we will obtain value of the critical radius:

$$\frac{d(\Delta G)}{dr} = \frac{d\left(\frac{4}{3}\pi r^3 \Delta G_V + 4\pi r^2 \gamma\right)}{dr} = 4\pi r^2 \Delta G_V + 8\pi r\gamma = 0 \qquad (3.9)$$

$$r^* = -\frac{2\gamma}{\Delta G_V} \qquad (3.10)$$

where ΔG_V is the negative value. Value of an energetic barrier for the stable clusters growth may be obtained by substituting relation (3.10) into Eq. (3.8):

$$\Delta G^* = \frac{16}{3}\pi \frac{\gamma^3}{(\Delta G_V)^2} \qquad (3.11)$$

Let us consider now the value of ΔG_V. According to the basic Gibbs equation, the free energy change depends on the work of the film growth:

$$d(\Delta G) = Vdp - SdT \qquad (3.12)$$

where V is the system's volume and S is the entropy of the system. Usually, a substrate temperature, T_s, is constant during the film growth process, therefore, an ideal gas law (see the Eq. (2.1)) will be a reasonable approximation in the form $pV = Nk_BT_s$. If we take into account that N/V is the density of atoms in the arriving flux of atoms, Eq. (3.12) transforms as follows:

$$d(\Delta G) = \frac{Nk_BT_s}{p}dp \qquad (3.13)$$

Let us designate Ω as a volume of the adsorbed atom (adatom) from the arriving flux. Then, with attachment of one atom to the cluster at temperature T_s, its free energy will be changed by the value

$\Delta G_V = \frac{\Delta G}{\Omega}$ and our equation will take the following form:

$$d(\Delta G_V) = \frac{k_B T_s}{\Omega}\frac{dp}{p}, \tag{3.14}$$

with the solution for this equation as

$$\Delta G_V = \frac{k_B T_s}{\Omega}\ln\frac{p}{p_{vp}} = \frac{k_B T_s}{\Omega}\ln\frac{J}{J_{vp}} = \frac{k_B T_s}{\Omega}\ln\zeta \tag{3.15}$$

Substituting the relation (3.15) into (3.10), we'll obtain the following value for the cluster's critical size:

$$r^* = \frac{2\Omega\gamma}{k_B T_s \ln\zeta}, \tag{3.16}$$

As shown, a thin film begins growing with the formation of small clusters of the critical size. The number of critical size clusters depends on the density of all possible places on the substrate suitable for the clusters' accommodation, n_s. In the first approximation, n_s represents the surface density of substrate's atoms with broken bonds. Thin films growth rate depends significantly on the energy of the interface on the unit area γ, of temperature of the substrate T_s and the binding energy ε_i between atoms of the coating material, that is atoms of the arriving flux J. If the substrate temperature and the supersaturation level are high enough, the critical size of the clusters decreases. Depending of the clusters critical size and relation between energy of interface and internal binding energy, the thin film growth will go in one of the following directions: monomolecular (monoatomic) growth or islands growth, as illustrated in Fig. 3.4.

The type of thin film growth may be observed using electron microscopy or atomic force microscopy techniques by definition of the so-called wetting angle α. This angle characterizes the energy relation. When $\alpha > 90^\circ$, we obtain the monomolecular or layer-by-layer type of growth called the Frank-Van der Merwe model. Here, the thin film has a good adhesion to the substrate and each layer of the film grows on the sublayer of the same material. When $\alpha < 90^\circ$, we obtain

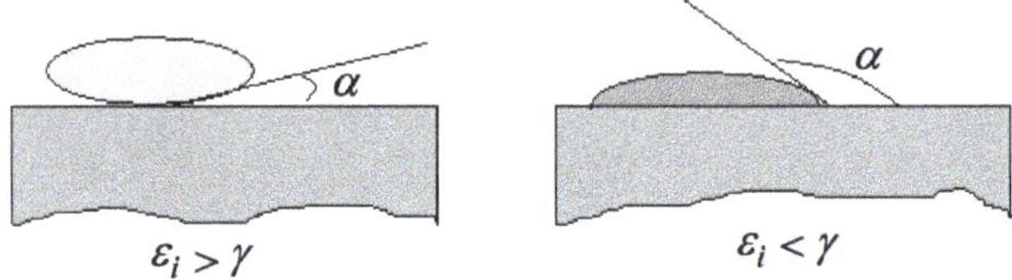

Fig. 3.4 Energy relation and the wetting angle.

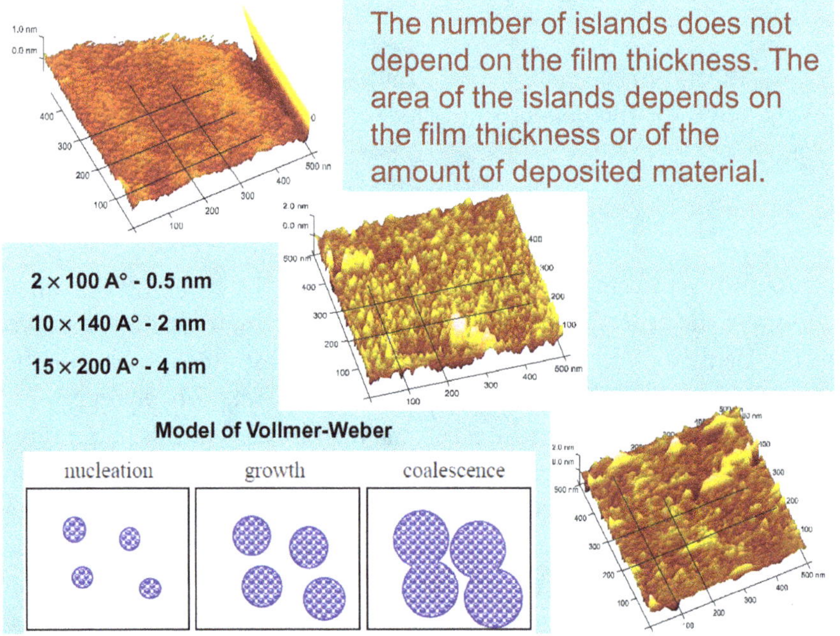

Fig. 3.5 Gold island films on a glass substrate.

the island growth on the thin film. Such type of growth called the Volmer–Weber model. In this case, after the formation of critical dimensions, clusters continue to grow and adsorb all other arriving atoms without formation of additional clusters up to the coalescence process. These films have poor adhesion to the substrate and require additional activation of the substrate to obtain good adhesion. Figure 3.5 illustrates the growth of very thin gold island films deposited by thermal evaporation on the glass substrate and observed using atomic force microscope. Gold thin films grow according to the Volmer–Weber model and they have poor adherence to glass.

To increase adhesion, the chromium sublayer is usually used before deposition of the gold film such as the chromium that grows by the Frank-Van der Merwe model and has good adhesion. As shown in Fig. 3.5, the gold thin films represent a constant number of clusters which grow in size after nucleation maintaining the same number of clusters but with additional coating material. Dimensions of the islands, height × diameter and average thickness of the films, are presented in the left side of Fig. 3.5.

Evidently, there is a third model of the thin films growth, called the Stranski–Krastanov model which enables the simultaneous growth of both layers and islands.

If the system is in the non-equilibrium state, natural forces appear always to bring the system into the steady state. Due to inhomogeneity of the material's composition or non-equity of temperature, mass-transfer effects also appear. These processes decrease the free energy of the system. They include the following effects: phase transitions, recrystallization, mixture growth, oxidation, degradation of the systems (for example, delamination of the thin films with time), etc. A diffusion, i.e., moving of atoms due to the concentration difference, is also the same natural process leading to the free energy decrease and the system equilibrium. Atoms arriving to the substrate during the thin film growth process create an elevated concentration of the novel material on the substrate surface. These atoms also have an additional energy, therefore, they begin moving on the substrate surface and the moving mechanism is called diffusion. Surface diffusion is a process involving the motion of adsorbed atoms (adatoms), molecules and atomic clusters on the solid substrate surface.

The diffusion process may be phenomenologically described using the well-known Fick's laws. Equations for the first and second Fick's laws in the one-dimensional form are as follows:

$$J = -D\frac{\partial C}{\partial x} - \text{First Law}; \quad \frac{\partial C}{\partial t} = D\frac{\partial^2 C}{\partial x^2} - \text{Second Law} \tag{3.17}$$

where $C, \left[\frac{\text{atoms}}{\text{cm}^3}\right]$, is the concentration of atoms, $J, \left[\frac{\text{atoms}}{\text{cm}^2\text{s}}\right]$, is the atomic flux moving from the high-concentration region to the low-concentration direction and D is the diffusion coefficient obeying the Arrhenius Law, as shown in the following relation:

$$D = D_0 e^{-\frac{E_A}{k_B T}} \tag{3.18}$$

Here, E_A is called an energy of the diffusion activation and D_0 is called the Debye coefficient. The joint solution of Fick's equations allows us to obtain the arriving atoms' distribution in space and time. Figure 3.6 illustrates the diffusion process on the substrate (a) and the diffusion mechanism. Arriving atoms begin to move from a place with higher concentration in the direction of lower concentration as shown in Fig. 3.6(a). Arrived atoms can move for a distance equal to the lattice parameter a due to their thermal oscillation with a frequency ν_0 if their energy will prevail the activation diffusion energy E_A in a casual way.

From the point of view of the atomistic theory of diffusion, a Debye coefficient depends on the frequency of oscillation related with the environment temperature and the lattice parameter of the substrate:

$$D_0 = a^2 \nu_0 = a^2 \frac{3}{2} \frac{k_B \theta_D}{h} \left[\frac{\text{cm}^2}{\text{s}} \right] \tag{3.19}$$

where θ_D is the Debye temperature for the material characterizing the crystal's higher normal mode of vibration, i.e., an oscillation in which all particles move with the same frequency and phase. This parameter

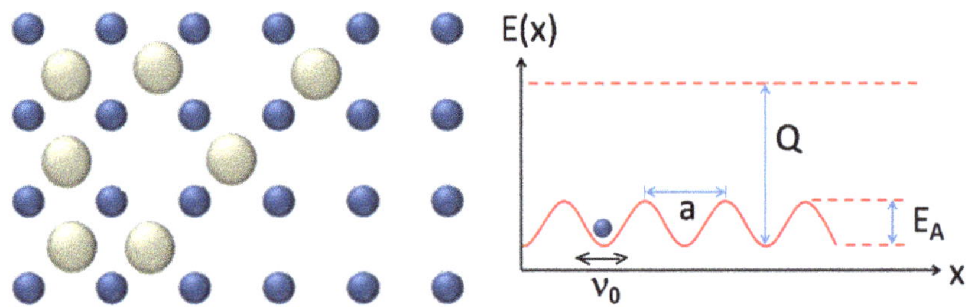

Fig. 3.6 An arriving atom's behavior on the substrate: (a) the diffusion process, (b) atom's oscillation and diffusion mechanism.

was measured for pure elements and compounds and may be found in the reference literature. Usually, the diffusion process is very slow, however, in conditions of the thin films growth it goes as fast as it occurs on the surface, and it depends on the arriving particles' energy and the substrate's temperature.

Physical vacuum deposition (PVD) is a family of processes used to deposit layers of material atom by atom or molecule by molecule on a solid surface. These processes operate at pressures well below atmospheric pressure (i.e., in vacuum). The thin deposited layers can range from a thickness of one atom up to approximately one micron. In general, all PVD methods may be divided into three big groups: thermal vacuum evaporation, sputtering (ion plasma deposition) and laser ablation. Here, we will focus on the first two methods, thermal vacuum evaporation and sputtering.

Both deposition methods, thermal evaporation and sputtering, mean dividing the precursor materials on separated atoms or molecules, creating a flux of these separated atoms directed to the substrate, where a thin film should be grown, and the actual thin film growth based on the phase transition process: steams-to-thin film. Main difference of these methods is in the energy of flowing particles (atoms or molecules), temperature of the process and the gas environment pressure and composition.

3.2 Vacuum thermal evaporation

Vacuum thermal evaporation method consists of heating a required material up to the evaporation temperature with phase transfer from the solid or liquid state to the gas phase. To provide this process realization, we need to put the precursor material in the evaporator and to heat it. There are two possible ways for this process: A direct resistive heating when the current flows through the metal evaporator or an indirect heating of the evaporator containing the material to be evaporated. In both cases, the evaporator should be heated. Usually, the so-called refractory metals such as tantalum (Ta), molybdenum (Mo) and tungsten (W) are used for preparation of the metal evaporators. As is known, tungsten is the metal that has the highest melting

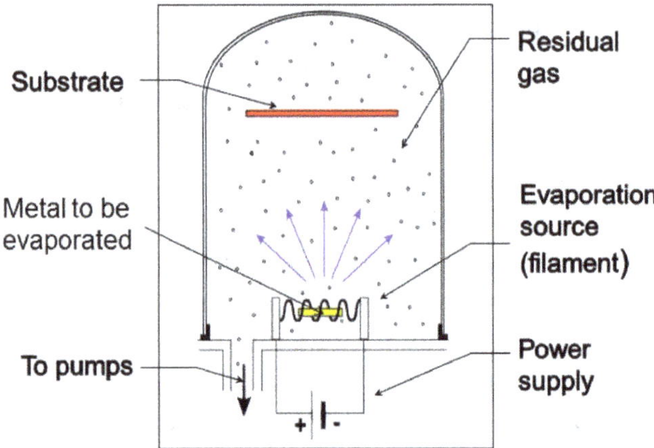

Fig. 3.7 Thermal evaporation setup.

temperature (3410°C) from all known elements. If we'll consider the temperature as a measure of the kinetic energy, the W melting point relates to approximately 30% of one electronvolt. Therefore, the maximum kinetic energy of evaporated particles, atoms or molecules, does not exceed this limit, 0.3 eV.

Thin films growth by thermal evaporation method may be presented as a sequence of the following three processes: materials decomposition with formation of a flux of atoms or molecules, mass-transfer of the atomic flux to the substrate, and a phase transition gas-to-solid during the thin film growth. Each process obeys the different physical phenomena and may be described by different parameters. Also, each of these three processes are separated in space. Figure 3.7 illustrates a thermal evaporation process.

As shown in Fig. 3.7, each process occurs in its place. **First process** is the thermal decomposition of materials under high-temperature heating. It occurs at the evaporator, which can be presented as a spiral filament built from tungsten, for example. The metal which can wet this filament, for example aluminum, should be arranged in the filament to be heated. Under heating by a high current flowing through the filament, aluminum transforms to liquid which wets the evaporator and transits into the gas phase (evaporates). During the process, a part of the evaporator reacts with the aluminum with

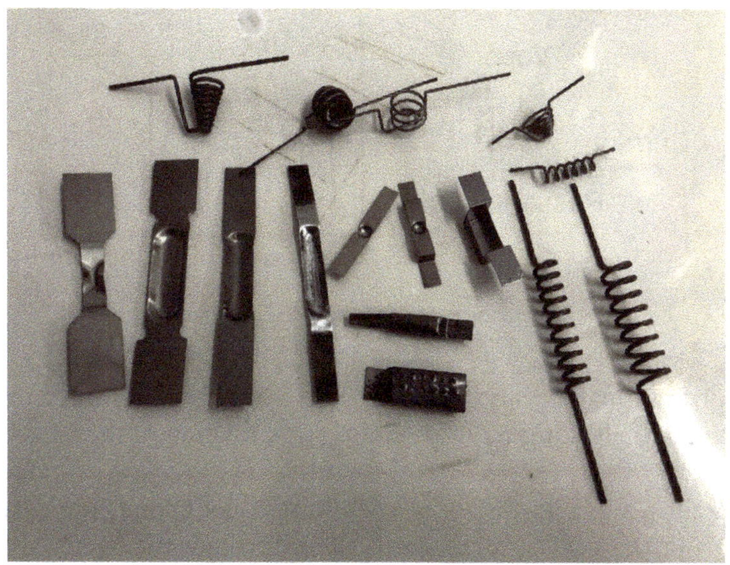

Fig. 3.8 Various types of resistive-heated evaporators.

appearance of the alloy of Al-W, which has a lower melting point than the tungsten. So, the number of evaporation processes provided with the same evaporator is limited. Evidently, an evaporator may be used with one evaporating material only. For each material, the specific evaporator should be chosen. Figure 3.8 presents various types of resistive-heated evaporators.

Evaporators, presented in Fig. 3.8, are of various types: W spiral filaments and baskets, W and Mo boats, Ta boxes open and closed with the perforated caps. For example, the Cu and Au thin films may be deposited by evaporation of the metal from W baskets. These metals, Cu and Au, do not wet the tungsten and after heating up to required temperature, they transit to the liquid phase shaped in the sphere form from which they evaporate. The closed Ta boxes are used for evaporation of powder materials. The perforated caps provide evaporation of atoms or molecules only and protect the substrate from big pieces of evaporated material.

When we begin heating a material in the evaporator, we bring it to an energy which swings the atoms of the material. Therefore, a part of atoms obtain casually enough energy to leave the material.

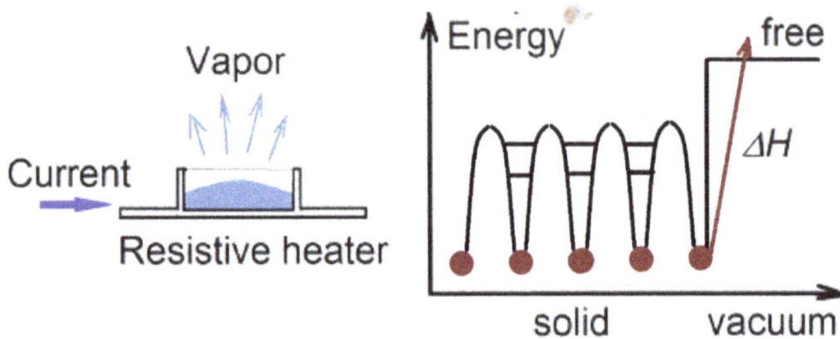

Fig. 3.9 Schematic presentation of the evaporation process.

This process may be illustrated by Fig. 3.9, where a parameter ΔH represents the heating to evaporation (an enthalpy) of the considered material.

As shown in Fig. 3.9, a resistive heater, containing the evaporation material, heats by a current flowing through the heater and the material evaporates. This process is illustrated by an energetic diagram when atoms of the evaporating material may leave it due to heating. In 1882, Dr. Heinrich Hertz provided the first experiments for various materials' evaporation and particularly for mercury. On the basis of these experiments, he has published an empirical relation for the flux of evaporated atoms:

$$\Phi_e = \frac{\alpha_e N_A (p_e - p_{\text{res}})}{\sqrt{2\pi MRT}} \tag{3.20}$$

where α_e is an evaporation coefficient, p_e is the evaporation material pressure dependent on the evaporator's temperature, p_{res} is the residual pressure, M is the atomic mass in g/mole and T is the temperature of the evaporation material. This relation looks like relation (2.13). If we assume that $\alpha_e = 1$ and $p_{\text{res}} = 0$, then the evaporation rate will be maximum. If the pressure of evaporated material is measured in Torr, the relation (3.20) transforms into the equation for the mass evaporation rate:

$$\Phi_e = 5.84 \cdot 10^{-2} \sqrt{\frac{M}{T}} p_e \left[\frac{\text{g}}{\text{cm}^2\text{s}}\right] \tag{3.21}$$

Let us consider Fig. 3.9 once more. As shown, to liberate an atom from the precursor material, we need to bring it an energy that is high enough, no less than ΔH. The evaporation in the phase conversion process is non-equilibrium due to the constant power supply. Therefore, the general equation for the Gibbs energy of the system will also be non-equilibrium:

$$G = H - TS - pV \tag{3.22}$$

For further consideration, we may divide our evaporation process into little intervals within which one can consider the evaporation process as an equilibrium process. Thus, we can reflect the change of the Gibbs energy as zero inside these short intervals:

$$0 = \Delta H - dT\Delta S - dp\Delta V \tag{3.23}$$

Here, we can assume that $\Delta H = 0$ due to our little intervals. If we designate the volume variation through the phase transition steam-condensate as $\Delta V = Vv - Vc$, where Vc is the condensate volume and Vv is the volume of the formed steam and $\Delta S = -\Delta H/T$, by definition, where ΔH is the constant and known value for each material, we obtain the following:

$$\frac{dp}{dT} = -\frac{\Delta S}{\Delta V} = \frac{\Delta H}{T\Delta V} \tag{3.24}$$

As such a volume of the condensate is very small, $Vc << Vv$, and substituting the ideal gas approximation $Vv = RT/p$ into Eq. (3.24), we obtain a relation describing the evaporation process:

$$\frac{dp}{dT} = \frac{p\Delta H}{RT^2} \Rightarrow \frac{dp}{p} = \frac{\Delta H}{R}\frac{dT}{T^2} \Rightarrow p = p_0 e^{-\frac{\Delta H}{RT}}, \text{where } p_0 = \text{const.} \tag{3.25}$$

As shown, the vapor pressure grows exponentially during temperature increases. It is generally believed that the evaporation pressure is approximately $p_e = 10^{-2}$ Torr for various materials. The evaporation temperature may be calculated from Eq. (3.25) or

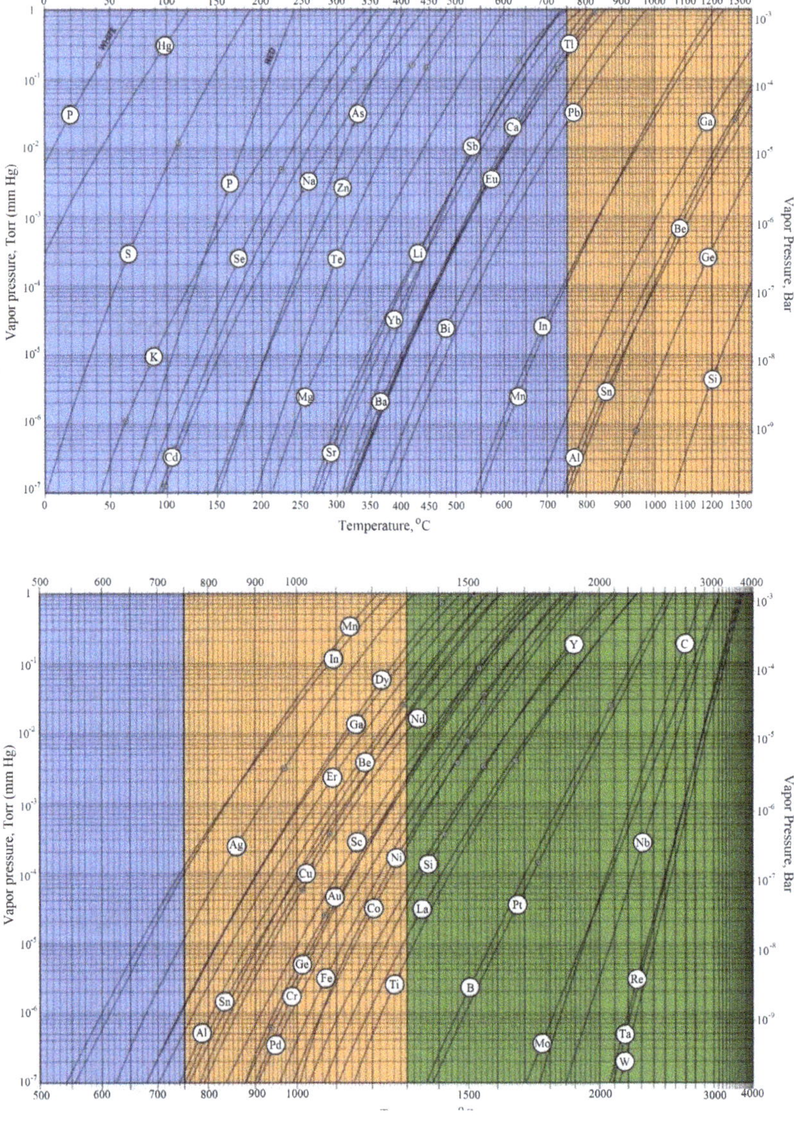

Fig. 3.10 Dependence of the vapor pressure on temperature of the materials.

obtained from diagrams measured for different elements and presented in Fig. 3.10. Evaporation temperatures can reach high values, so there are metals that cannot be evaporated. Thin films from these metals may be prepared by other physical deposition methods.

For preparation of complex thin films consisting of alloys, we need to take into account that various components will evaporate at different rates from the alloys at the constant temperature. So, in this case, it is desirable to use two or more different evaporators, each one for the suitable component. Another way is to calculate a required relation of components in the future thin films and evaporate all materials from the evaporator. This method requires the past-deposition sintering process at high temperature to obtain required composition and structure of the deposited thin film. Evaporation process of multicomponent materials obeys Raoult's law. It states that the partial pressure of each component of an ideal mixture is equal to the vapor pressure of the pure component multiplied by its mole fraction in the mixture:

$$p_i = x_i p_i^* \tag{3.26}$$

where $i = 1, 2, \ldots$ is a number of the component, x_i is the mole fraction of the component and p_i^* is the equilibrium vacuum pressure of the pure component. Accordingly, for a two-component compound evaporated from the one evaporator, we can write a relation of the evaporated fluxes as follows:

$$\frac{\Phi_1}{\Phi_2} = \frac{x_1}{x_2}\frac{p_1^*}{p_2^*}\sqrt{\frac{M_2}{M_1}} \tag{3.27}$$

Using this equation, we can approximately define amounts of the components in the evaporator for deposition of metal alloy films with the predicted composition. It should be noted that this calculation is not precise due to changes in the deposition parameters through the process.

Evaporation of semiconductor or dielectric compounds such as oxides or nitrides has its own characteristics. There are compounds which can evaporate without dissociation and decomposition and there are other materials which decompose during the evaporation process. These materials require specific techniques for deposition of the stoichiometric compound thin films. Table 3.1 represents examples of the evaporation conditions for various materials.

Table 3.1 Several features of compounds' evaporation.

Reaction type	Chemical reaction	Examples	Comments
Evaporation without dissociation	MX(s or l) → MX(g)	SiO, B_2O_3, GeO, AlN CaF_2, MgF_2	Thin film preserves stoichiometry
Decomposition	MX(s) → M(s) + $1/2X_2(g)$ MX(s) → M(l) + $1/nX_n(g)$	Ag_2S, Ag_2Se III–V semi conductors	Separated sources are required for deposition
Evaporation with dissociation			Thin films are metal-rich
Chalcogenides	MX(s) → M(g) + $1/2X_2(g)$ X = S, Se, Te	CdS, CdSe, CdTe	Separated sources are required for deposition
Oxides	$MO_2(s)$ → MO(g) + $1/2O_2(g)$	SiO_2, GeO_2, TiO_2, SnO_2, ZrO_2	Metal-rich discolored films, oxygen rich atmosphere required

Note: M = metal, X = non-metal, s = solid, l = liquid, g = gas phase.

The not-required decomposition of oxides or nitrides during evaporation may be compensated for by reactive evaporation in an oxygen or nitrogen environment.

The **second part** of the evaporation process occurs in the space between evaporator and substrate. Evaporated atoms or molecules should be transferred to the substrate. Now, to keep the stoichiometry of the growing thin films, we need to provide transfer of the evaporated materials in pure conditions. Atoms must jump from evaporator up to substrate without collision with the residual gas atoms. Therefore, enough high vacuum should be reached in the vacuum chamber. An approximation formula (2.11) enables us to estimate the mean free path of the residual gas,

$$\lambda = \frac{5 \cdot 10^{-3}}{p}, \text{ where } p = [\text{Torr}] \text{ and } \lambda = [\text{cm}] \qquad (3.28)$$

To provide enough pure conditions and decrease the influence of the residual gas on the properties of the growing thin film, a distance between the evaporator, the source of the precursor material and the substrate should be more than 10λ. By this, one can defines the minimum vacuum (maximum pressure) required for the evaporation process. For example, if we have the vacuum thermal evaporation setup with a distance between source and substrate equal to h = 15 cm, the maximum pressure permissible in the vacuum chamber will be $p = \frac{5 \cdot 10^{-3}}{10 \cdot 15} \approx 3 \cdot 10^{-5}$ Torr. Higher vacuum will provide more pure conditions.

Sometimes, the evaporation process should be provided in an atmosphere of reactive gases such as oxygen, nitrogen or others. Such processes were mentioned in Table 3.1. In these cases, a vacuum chamber should be pumped up to residual pressure and only after should the reactive gas be inserted in the vacuum chamber, in the substrate region desirable. Figure 3.11 illustrates a schematic view of the system for vacuum evaporation in plasma of evaporated steams or a reactive gas environment.

The system presented in Fig. 3.11 enables providing thermal evaporation process and influence on the grown thin film properties by

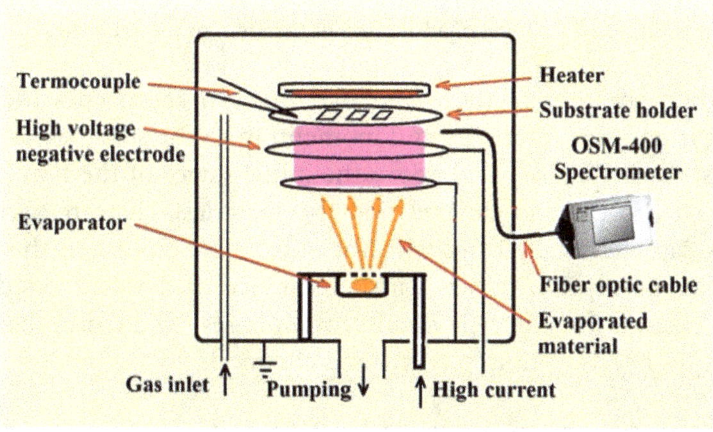

Fig. 3.11 Schematic view of the ion-plating setup enabling reactive evaporation in plasma.

application of the plasma of glow discharge in the steams of evaporated materials. Also, the evaporation process may be provided in the reactive gas plasma environment, which enables to grow oxides, nitrides, sulfides and other complex compositions from evaporated metal. In this system, evaporated material goes through the ring electrodes producing the DC or RF plasma from the independent supply.

The **third part** of the evaporation process occurs on the substrate surface. We have discussed above the beginning of the film growth. However, the first requirement for the film is the adhesion of the adsorbed film to the substrate. Let us consider a model of the processes occurring when an atom approaches the substrate surface, as shown in Fig. 3.12.

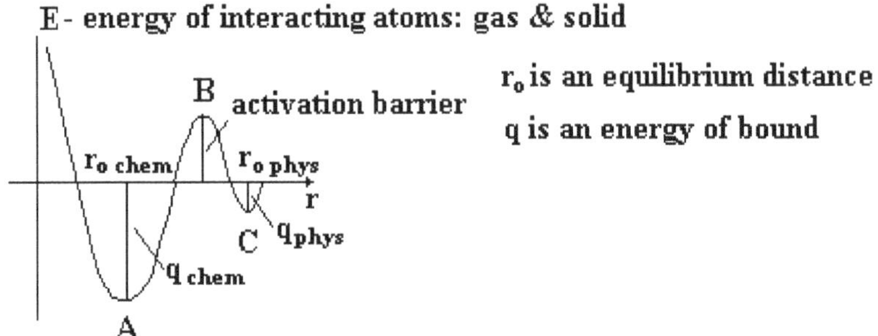

Fig. 3.12 Model of adsorption of the atom to the substrate.

When an atom of the evaporated material approaches the substrate, it transfers its energy and momentum to the substrate and can be adsorbed. We discussed above the initial stages of the film incipience. Due to the energies' relation and materials' character, arriving atoms can chemically react with the substrate's atoms. In this case, chemical adsorption occurs with appearance of covalent bonds. In other cases, a physical type of adsorption arises. Therefore, there are two types of adsorption, physical and chemical:

- Physical adsorption is defined by the van der Waals and electrical polarization forces;
- Chemical adsorption is defined by creation of the covalent bonds or sharing electrons from neighboring atoms of the substrate and the growing thin film.

Both adsorption types are different by the energy of bonds: $\mathbf{q_{phys}}$ **~0.01–0.1 eV**, $\mathbf{q_{chem}}$ **~ 1 eV**. Due to the energy difference between two adsorption processes, adhesive force will also be different. So, to obtain a good (high) adhesion, the film should be grown according to a two-dimensional model (see Fig. 3.4) and it is desirable for the creation of chemical bonds between the substrate and the coating.

Another problem which appears when we begin to grow thin films by the thermal evaporation method is a thickness uniformity of the grown film. An evaporation source may have various forms, however, for simplicity, we can present it as a point source. A simplified model of the evaporation process is presented in Fig. 3.13.

Let us assume that one material with density ρ and mass M is evaporated from the point source which is arranged at a distance h from the substrate with the diameter $2r$. Evaporation from the point source occurs in all directions equally. Therefore, after full evaporation of the material, a thin film with the thickness d_0 will coat all inner surfaces of the sphere with a radius h. This may be described by the following equation:

$$M = 2\pi\rho d_0 h^2 \tag{3.29}$$

where $2\pi h^2$ represents the internal surface of the sphere with the radius h. Therefore, the thickness of the film in the center of the round flat substrate will be equal to

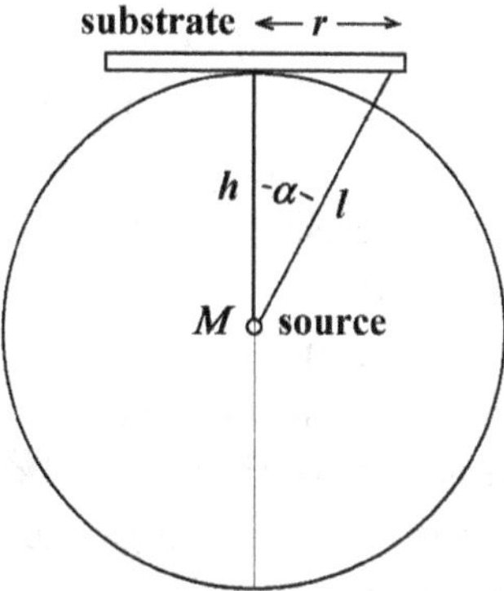

Fig. 3.13 Schematic view of the evaporation process from the point source.

$$d_0 = \frac{M}{2\pi\rho h^2} \tag{3.30}$$

However, the thickness distribution on the substrate surface is also an important parameter of the grown film. As can be seen in Fig. 3.13, the distance l is more than the distance h and it may be defined by a simple geometric relation:

$$l = \sqrt{h^2 + r^2} = \frac{h}{\cos\alpha} \tag{3.31}$$

Therefore, the thickness of the film will be different on the substrate surfaces depending on the distance from the substrate center. The thickness distribution on the substrate surface will be described by the following equation called "Cosine's law":

$$d_0 = \frac{M}{2\pi\rho h^2}\cos\alpha = \frac{M}{2\pi\rho h^2}\frac{h}{\sqrt{h^2 + r^2}} \tag{3.32}$$

Evidently, we considered only the simplest case of an evaporation system arrangement. In reality, the evaporation sources are not point-like and have dimensions on the order of a few millimeters. However, the presented analysis is enough for prediction of the deposited film thickness and its distribution on the substrate surface. There are many variations for improving the thickness distribution, for example by increasing the distance between evaporation source and a substrate, rotation of the substrate and suitable arrangement of the substrates inside a vacuum chamber, as shown in Fig. 3.14.

Presented in Fig. 3.14 are methods that are applied in the thermal evaporation systems, however, usually only the flat rotation method is used as it is the simplest. Precision mechanic works are required for all other improvement methods, which significantly increases the price of the evaporation equipment.

Thus, we considered the thermal evaporation systems and concluded that the thickness distribution and all other properties of the grown thin film depend on several things: geometry of the vacuum chamber and internal installation, temperature and vapor pressure in

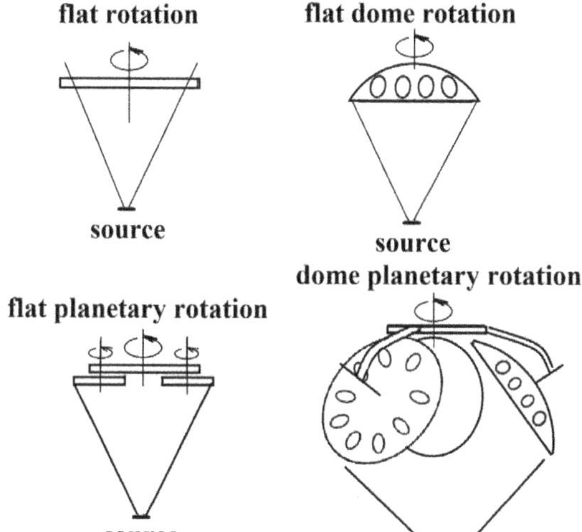

Fig. 3.14 Various methods for the improvement of thickness distribution.

the vacuum chamber, properties of applied materials. Three different temperatures define this thin film deposition process: temperature of the evaporation source, temperature of the substrate and temperature of the vacuum chamber walls. Relation between these temperatures should provide the evaporant steams pressure enough for evaporation, the required ratio between energies of the evaporant atoms and the substrate, and a low mobility of the residual gas atoms near the vacuum chamber walls. Such a relation may be described as follows:

$$T_{\text{source}} \approx T_{\text{evaporant}} \gg T_{\text{substrate}} > T_{\text{chamber}} = T_{\text{residual gas}} \tag{3.33}$$

Above, we considered the vacuum evaporation process using the direct heated evaporation sources only. However, for thin films deposition by evaporation, there are other techniques, such as laser flash evaporation, evaporation using an electron evaporator or an electron gun, evaporation using the Knudsen cells or the effusion cells as evaporator, etc. All these methods are out of the scope of this book and here we describe the deposition and evaluation methods applicable in the training laboratory and available for students.

3.3 Plasma of the gas discharge

We know well that substances can be in three phase states: solid, liquid, gas. These phase states may be defined as follows: the solid state — the matter keeps a form and a volume; the liquid state — the matter keeps a volume, but does not keep a form; the gas state — the matter does not keep a volume or a form. Transition from the solid state to the liquid and to the gas state requires energy to rock and break the bonds between atoms or molecules. If we are to take a gas at normal pressure, close it in the confined volume and attach an energy to this gas, there is a possibility to release electrons from a part of the gas atoms. Such a mixture from three components of gases: neutral molecules, ions and electrons, is called plasma. Due to random movement of gas molecules, electrons and ions, casual collisions occur in the plasma volume. Collisions may be of several types. For example, a collision of an electron with an argon atom may excite the atom to a higher energy level without ionization: $e + Ar \Rightarrow Ar^* + e$ (excitation process), corresponding to the change of state $3p^6 \rightarrow 4p^5 4s^1$ for the valence electrons. This metastable atom can then emit a photon and return to a lower energy state $e + Ar \Rightarrow Ar^* + e \Rightarrow Ar + e + h\nu$ (relaxation process). The radiation is usually in the visible or ultraviolet spectrum. Each gas is known to possess an individual color at the plasma state. This color is explained by both radiation processes: relaxation and recombination. Each act of recombination is accompanied by an act of radiation. Other processes are ionization acts of the following types: $e + Ar \Rightarrow Ar^-$ or $e + Ar \Rightarrow 2e + Ar^+$. The lifetime of these ions is defined by the gas pressure and other external acting factors, for example by applied external electric field.

The mixture of mentioned above plasma components is a dynamical substance, many different processes occur in plasma and we define two of their general properties: plasma has a measurable conductivity and plasma has a characteristic color. Table 3.2 represents characteristic colors appearing by plasma in several gases. These properties are due to the presence of free electrons and ions. The charged carriers provide conductivity of plasma and the recombination processes define the characteristic colors. It is of interest that a current in

Table 3.2 Characteristic colors of various gases.

Gas	Negative glow	Positive column
Argon	Blue	Violet
Carbon tetrachloride	Light green	Whitish green
Carbon monoxide	Greenish white	White
Carbon dioxide	Blue	White
C_2H_5OH	—	Whitish
Cadmium	Red	Greenish blue
Hydrogen	Light blue	Pink
Mercury	Whitish yellow	Blue green
Potassium	Green	Green
Krypton	Violet	Yellow pink
Air	Blue	Reddish
Nitrogen	Blue	Red-yellow
Sodium	Whitish	Yellow
Oxygen	Yellowish white	Lemon yellow with pink core
Thallium	Green	Green
Xenon	Pale blue	Blue violet

semiconductors and metals can be called plasma as well, as such a current represents the movement of charged carriers. Moreover, a movement of liquids containing charged particles, for example blood, also can be called plasma. Later, we will show that many processes in plasma and in the semiconductors are described by the same equations.

In general, gases are dielectrics. To release an electron from a gas molecule, it is needed to bring an ionization energy to the molecule. There are several methods to bring energy to the gas molecules. This may be heating, application of the static electrical field, application of the radio-frequency electromagnetic field, gamma-rays, etc. Let us consider a process of plasma formation. For this goal, we will take a system consisting of two plane parallel electrodes which are in a closed space, for example in a vacuum chamber, and we can control the gas pressure in this space and the voltage between electrodes. Figure 3.15

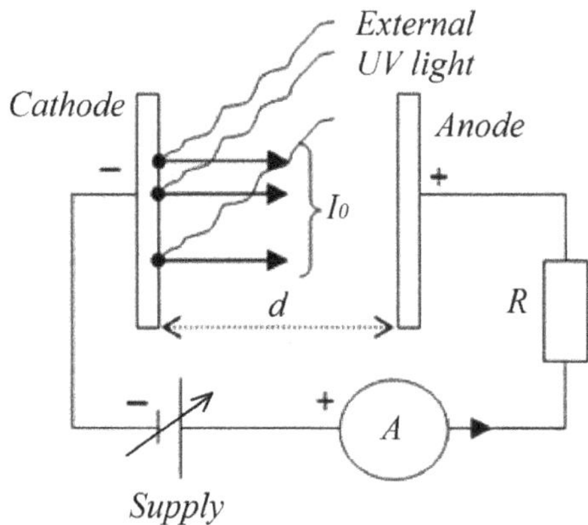

Fig. 3.15 A schematic system for providing experiments with plasma.

represents the schematic system for our analysis. Evidently, there are many factors influencing this system: gas pressure, distance between electrodes, environment temperature, materials and quality of the surfaces of electrodes, quantity of initial charged carriers in the gas volume, etc.

Assume that in the system shown in Fig. 3.15 the gas pressure is equal to p and the voltage V applied between electrodes produces in this space the electrical field $E = V/d$ where d is the distance between electrodes. Also, assume that the cathode is irradiated by ultraviolet light enough for emission of electrons from the cathode. Under influence of the electrical field, electrons will move with a velocity v to the anode gaining a kinetic energy while driving, according to the energy conservation equation, which looks as follows:

$$qEx = \frac{1}{2}mv^2 \tag{3.34}$$

where x is the distance of the electron from the cathode. Electrons will collide with gas molecules on the way to the anode. If their energy is high enough for ionization (for example, 12.2 eV is enough for ionization of the nitrogen molecules and 15.5 eV — for oxygen), novel electrons will be created due to collisions. These electrons will be attached to the initial electron flow creating by this way the avalanche process and increasing the current between electrodes. This increasing of the current causes a breakdown of the gap between anode and cathode.

Assume that the number of electrons emitted from the cathode under UV light influence is equal to n_0 and due to impact of ionization, one more electron is created for each collision of an electron with the gas molecule. Also assume that the electron produces in average α ionization acts on the distance 1 cm while moving to the anode. This number, α, dependent on the gas pressure p and a relation between the electrical filed and the gas pressure E/p, is called the first ionization coefficient of Townsend. Thus, on the distance x from the cathode, a number of electrons will be n_x and on the distance dx, the number dn_x of electrons will be created:

$$dn_x = an_x dx \tag{3.35}$$

From this equation, the number of electrons which reached the anode will be (for $x = d$):

$$n_d = n_0 e^{\alpha d} \tag{3.36}$$

Accordingly, an average current through the gap will be equal to $I = I_0 e^{ad}$. The number of electrons created by one electron due to ionization processes will be as follows:

$$\frac{n_d - n_0}{n_0} = e^{\alpha d} - 1 \tag{3.37}$$

The ions produced due to ionization acts move to the cathode. Due to collisions with electrons, a part of them recombine and emit photons, however, a part of ions having enough momentum and kinetic energy impact the cathode producing by this way the secondary electron emission. The secondary electron emission from a metal cathode is described by the second emission coefficient of Townsend, γ, which also depends on the gas pressure p and a relation between the electrical filed and the gas pressure E/p. If we designate a number of electrons emitted by the secondary emission mechanism by n'_0, the total amount of emitted electrons from the cathode will be $n_0'' = n_0 + n_0'$. Therefore, the total amount of electrons reaching the anode will be

$$n = n_0'' e^{\alpha d} = \left(n_0 + n_0'\right) e^{\alpha d}, \text{where } n_0' = \gamma\left[n - \left(n_0 + n_0'\right)\right] \quad (3.38)$$

Therefore, a number of electrons reaching the anode and a current through the gap anode–cathode will be equal to the following:

$$n = \frac{n_0 e^{\alpha d}}{1 - \gamma\left(e^{\alpha d} - 1\right)} \text{ and } i = \frac{I_0 e^{\alpha d}}{1 - \gamma\left(e^{\alpha d} - 1\right)} \quad (3.39)$$

A sharp growth of the current through the gap between electrodes will appear when the denominator in Eq. (3.39) will tend to zero. So, for a defined distance between electrode, d_s, one can write the Townsend criterion as follows:

$$1 - \gamma\left(e^{\alpha d} - 1\right) = 0 \text{ or } d_s = \frac{1}{\alpha}\ln\left(1 + \frac{1}{\gamma}\right) \quad (3.40)$$

According to the definition, the emission coefficients α and γ depend on the relation between the electrical field and the gas pressure E/p:

$$\frac{\alpha}{p} = f_1\left(\frac{E}{p}\right), \gamma = f_2\left(\frac{E}{p}\right), E = \frac{V}{d} \quad (3.41)$$

Substituting these relations into the Townsend criterion, we obtain the following equation relating a distance between electrodes and a breakdown voltage, $V = f(pd)$:

$$f_2\left(\frac{V}{pd}\right)\left[e^{pdf_1\left(\frac{V}{pd}\right)} - 1\right] = 1 \quad (3.42)$$

It was found experimentally that an ionization coefficient α depends exponentially on the gas pressure and the applied electrical field:

$$\alpha = pAe^{-\frac{Bp}{E}} \tag{3.43}$$

Substituting Eqs. (3.42) and (3.43) into the Townsend criterion will give the famous equation representing the Paschen law that gives the breakdown voltage in the gaps within the gas:

$$V_{Bd} = \frac{Bpd}{\ln(pdA) - \ln\left[\ln\left(\frac{1}{\gamma} + 1\right)\right]} \tag{3.44}$$

Here, the constant coefficients *A* and *B* are experimental coefficients characterizing the gas. Paschen's law enables to calculate a breakdown voltage between two electrodes in a gas as a function of the gas pressure and the distance between electrodes. Figure 3.16 represents

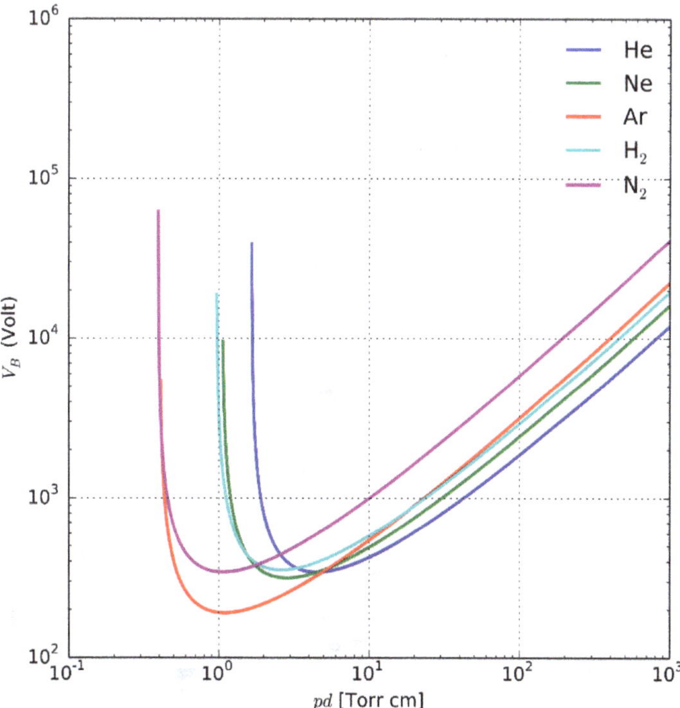

Fig. 3.16 The Paschen curves calculated for various gases.

the Paschen curves calculated for different gases using the expression (3.44) for the breakdown voltage as a function of the parameters A and B for the cathode prepared from the tungsten ($\gamma = 0.095$). It is of interest to note that according to the product of two parameters, distance between electrodes and gas pressure, the Paschen curve presents two possibilities of the electrical isolation of electrodes: vacuum isolation and high-pressure isolation. These possibilities are used in the industry, for example, for isolation of electrical wires on air environment using ceramic insulators.

After breakdown of the gas gap between cathode and anode, the self-sustained gas discharge appears in the vacuum chamber. This gas discharge consists of several glowing areas. A main plasma region called a positive column has approximately equal number of electrons and ions and a uniform electrical field strength along this region. Figure 3.17 illustrates schematically the diode glow discharge.

As shown in Fig. 3.17, a glow discharge contains a main plasma region with approximately equal number of charged particles. However, the electron velocity is very high compared with the ion velocity, which leads to local inhomogeneity in the plasma. Each inhomogeneity produces the free charges, which create electrical fields influencing these free charges. Free charged particles move randomly and create the charge inhomogeneity:

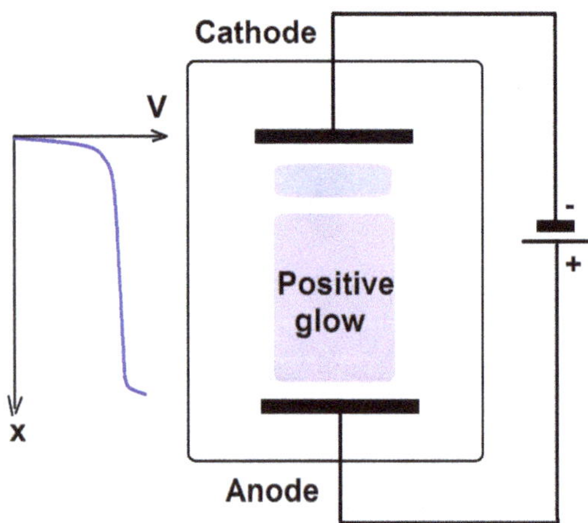

Fig. 3.17 The diode glow discharge in argon.

$$\rho = n_e(-q) + n_i(q) = q(n_i - n_e) \tag{3.45}$$

This charge difference produces an electrical field:

$$\nabla E = \frac{\rho}{\varepsilon_0} = \frac{q}{\varepsilon_0}(n_i - n_e) \tag{3.46}$$

That in the one-dimensional case presents Paschen's electrical field influencing the charge:

$$\frac{dE}{dx} = \frac{\rho}{\varepsilon_0} = \frac{q}{\varepsilon_0}(n_i - n_e) \Rightarrow E = \frac{q}{\varepsilon_0}(n_i - n_e)x = \frac{q}{\varepsilon_0}\Delta n \cdot x \tag{3.47}$$

The force influencing on the charged particle will be equal to:

$$F_q = \rho E = \frac{q^2}{\varepsilon_0}(\Delta n)^2 \cdot x \tag{3.48}$$

This force is proportional to the distance and acts in the reverse direction of charge division. One can say that this force plays the role of feedback (constructive) and it supports the plasma to the steady state. Also, destructive forces act in the plasma, for example the thermal motion forces, which may be described approximately by the following relation:

$$F_p \approx \frac{p}{x} = \frac{n_e T_e}{x} \tag{3.49}$$

where p is the pressure in the vacuum chamber, n_e is the concentration of ionized particles or electrons and T_e is the temperature of electrons. In general, $F_q >> F_p$ and plasma behaves as a dynamic net which seeks to maintain their structure.

The charged particles, ions and electrons, are randomly distributed inside the positive plasma column. The constant movement of these particles influences the plasma density. To describe the plasma behavior, we can use the differential form of the known Maxwell equations in vacuum:

$$\begin{cases} \nabla \cdot \boldsymbol{E} = \dfrac{\rho}{\varepsilon_0} \text{ (Gauss's law)} \\ \nabla \cdot \boldsymbol{H} = 0 \text{ (Gauss's law for magnetism)} \\ \nabla \times \boldsymbol{E} = -\mu_0 \dfrac{\partial \boldsymbol{H}}{\partial t} \text{ (Faraday's law for induction)} \\ \nabla \times \boldsymbol{H} = \varepsilon_0 \dfrac{\partial \boldsymbol{E}}{\partial t} + j \text{ (Ampere's circuital law)} \end{cases} \tag{3.50}$$

where $\boldsymbol{E}$ and $\boldsymbol{H}$ are electric and magnetic field vectors and $\mu_0 = 4\pi \cdot 10^{-7}$ H/m and $\varepsilon_0 \approx 8.85 \cdot 10^{-12}$ F/m are, respectively, the free space permeability and permittivity. The sources of the electrical and the magnetic fields are, respectively, the charge density ρ and the current density j. The charge density changes in a point can be obtained by taking a partial time derivative of the equation of Gauss law and substituting the expression into the equation of Ampere law:

$$\frac{\partial}{\partial t}(\varepsilon_0 \nabla \boldsymbol{E}) = \frac{\partial \rho}{\partial t} \tag{3.51}$$

$$\frac{\partial}{\partial t}(\varepsilon_0 \nabla \boldsymbol{E}) = \nabla\left(\varepsilon_0 \frac{\partial \boldsymbol{E}}{\partial t}\right) = \nabla(\nabla \times \boldsymbol{H} - j) = \nabla(\nabla \times \boldsymbol{H}) - \nabla j \tag{3.52}$$

Taking into account the known formula of the vector analysis, $\nabla(\nabla \times \boldsymbol{H}) = 0$, and substituting both of the last equations, we get the charge continuity equation in the form of the charge conservation law:

$$\nabla j + \frac{\partial \rho}{\partial t} = \nabla j + q \frac{\partial (n_i - n_e)}{\partial t} = 0 \tag{3.53}$$

This equation describes a relation between the current and charges distribution in a small part of plasma volume. Figure 3.18 illustrates such a small plasma volume with a current and the ionization-recombination processes occurring in plasma in the one-dimension form.

Here, in Fig. 3.18, A is the cross-section of an elementary volume of plasma with thickness dx, j_n or j_i are electron or ion current

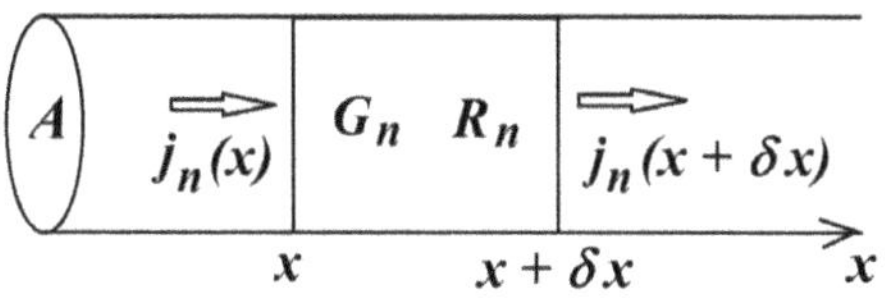

Fig. 3.18 The small plasma volume behavior.

densities and x is the moving direction. Considering the assumption of the weakly-ionized plasma and ionization-recombination processes, Eq. (3.53) can be written independently for each type of charged carriers as follows:

$$\frac{\partial n_i}{\partial t} \cdot A \cdot dx = \frac{A}{q}\left[j_i(x) - j_i(x + \delta x)\right] + G_i \cdot A \cdot dx - R_i \cdot A \cdot dx \quad (3.54)$$

$$\frac{\partial n_e}{\partial t} \cdot A \cdot dx = \frac{A}{q}\left[j_e(x) - j_e(x + \delta x)\right] + G_e \cdot A \cdot dx - R_e \cdot A \cdot dx \quad (3.55)$$

If we assume that the currents change continuously in the x direction (at least in the positive column region), we can write as follows:

$$J_i(x) - J_i(x + \delta x) = -\left(\frac{\partial J_i}{\partial x}\right) dx \quad (3.56)$$

Accordingly, Eqs. (3.54) and (3.55) can be rewritten as follows:

$$\frac{\partial n_i}{\partial t} = -\frac{1}{q_i}\frac{\partial J_i}{\partial x} + G_i - R_i \text{ and } \frac{\partial n_e}{\partial t} = -\frac{1}{q_i}\frac{\partial J_e}{\partial x} + G_e - R_e \quad (3.57)$$

In the single-ionized gas plasma the ion charge is equal to the elementary electron charge $q_i = q_e = q = 1.6 \cdot 10^{-19}$ C. Therefore, the ion current in the glow discharge is defined by both diffusion and electrical drifts:

$$J_i = J_{iD} + J_{iE} = q\mu_i n_i E - eD_i \frac{dn_i}{dx} \quad (3.58)$$

where μ_i is the ion mobility in the discharge, and D_i is the ion diffusivity.

Substituting (3.58) into (3.57), we arrive at the complete continuity equations:

$$\frac{\partial n_i}{\partial t} = -n_i \mu_i \frac{\partial E}{\partial x} - \mu_i E \frac{\partial n_i}{\partial x} + D_i \frac{\partial^2 n_i}{\partial x^2} + G_i - R_i \tag{3.59}$$

$$\frac{\partial n_e}{\partial t} = n_e \mu_e \frac{\partial E}{\partial x} + \mu_e E \frac{\partial n_e}{\partial x} + D_e \frac{\partial^2 n_e}{\partial x^2} + G_e - R_e \tag{3.60}$$

These equations describe the ion and electron concentrations' behavior in the gas discharge plasma. It should be noted that they are similar to the continuity equation for semiconductors and can be analyzed in a similar way.

The Gauss law claims that the electric field is created by the charge carrier's distribution. One can say that each charged particle has an electric potential φ dependent on the space coordinates. Thus, for single-ionized particles we have:

$$\varphi = \rho / r \tag{3.61}$$

where r is a coordinate in space and $\rho = q$ is the charge of the particle. Each particle in plasma is driven with thermal velocity in random directions, with distribution subject to Boltzmann's rule. The electrical field E is equal to the potential gradient and acts in the opposite direction:

$$\boldsymbol{E} = -\boldsymbol{\nabla} \varphi \tag{3.62}$$

Substituting Eq. (3.62) into the Gauss law yields the well-known Poisson equation:

$$\varepsilon_0 \boldsymbol{\nabla E} = \varepsilon_0 \boldsymbol{\nabla}(-\boldsymbol{\nabla} \varphi) = \varepsilon_0 \Delta \varphi = \rho \tag{3.63}$$

If we designate n_0 as the charged carriers' concentration on the distance ρ when the potential $\varphi = 0$, the charged particles' concentration will obey the Boltzmann concentration

$$n = n_0 e^{-\frac{q\varphi}{kT}} \Rightarrow \Delta\varphi = -\frac{q}{\varepsilon_0} n_0 e^{-\frac{q\varphi}{kT}} \tag{3.64}$$

To solve this equation, we simplify it, assuming the random deviations of the momentary concentrations and potential fluctuations. Based on these assumptions, we can develop the exponential function to the Taylor series and stop after the second term, thus linearizing the right-hand side in Eq. (3.64):

$$\Delta\varphi = -\frac{n_0 q^2}{\varepsilon_0 kT}\varphi \tag{3.65}$$

For a spherical symmetrical distribution of potential near the charged particle, we obtain the solution in the following form:

$$\varphi = \frac{C}{r} e^{-\frac{r}{\sqrt{\frac{\varepsilon_0 kT}{n_0 e^2}}}} \tag{3.66}$$

where C is an integration constant that approximates q (see prediction (3.61)) and the screening length L_D, called the Debye length, looks as follows:

$$L_D \equiv \lambda_D = \sqrt{\frac{\varepsilon_0 kT}{n_0 e^2}} \tag{3.67}$$

The Debye length is a typical length such that a small deviation in the charge density from the equilibrium within this length is relaxed or screened (in other words, is no longer felt). A characteristic feature of plasma is its ability to screen out an electrical charge. A potential disturbance in plasma will attract particles of the opposite charge. This cloud of charge provides screening from the rest of the plasma. This phenomenon is referred to as Debye screening. Due to Debye screening and electrical field affecting on the charged particles, the positive column in plasma represents a quasi-neutral stable system. If we designate N_D as a number of particles inside the Debye's sphere, $N_D = n\frac{4}{3}\pi\lambda_D^3, N_D < 1$ means that the plasma parameters will be influenced by collisions between particles and $ND >> 1$ means that the properties of the plasma will be defined by the joint behavior of the particles.

Electrons and ions move with thermal velocities just like neutral molecules do. An electron mass is much smaller than that of an ion. Therefore, staying within the cold plasma approximation, one can suppose that the ions rest while the electrons move among them. Spontaneous motion results in a small deviation of the space charges and creates charge divisions in the plasma volume. Therefore, according to the charge conservation law (Eq. (3.53)) and assuming that only electrons are responsible for the electric current, one can write as follows:

$$\frac{\partial \rho}{\partial t} = -\nabla \boldsymbol{j} = -\nabla\left(-nq\boldsymbol{v}\right) = nq\nabla\boldsymbol{v} \tag{3.68}$$

where v is the electron velocity. Charge division creates an electrical field E that affects the electron motion:

$$m\frac{d\boldsymbol{v}}{dt} = -q\boldsymbol{E} \tag{3.69}$$

Substituting Eq. (3.69) into Eq. (3.68), and differentiating, we get

$$\frac{\partial^2 \rho}{\partial t^2} = -\frac{nq^2}{m}\nabla\boldsymbol{E} \tag{3.70}$$

Considering the equation of the Gauss law and assuming that the full derivative is equal to the partial one, we obtain the following equation to describe the casual motion of plasma charge density:

$$\frac{d^2 \rho}{dt^2} = -\frac{nq^2}{\varepsilon_0 m}\rho = -\omega_0^2 \rho \tag{3.71}$$

This equation describes the natural sinusoidal oscillation of the electron cloud with respect to the ion cloud with the natural frequency ω_0, called the Langmuir frequency. This frequency is the self-frequency of plasma; it is defined only by the following plasma concentration:

$$\omega_0 = \sqrt{\frac{nq^2}{\varepsilon_0 m}} \tag{3.72}$$

For example, for electrons, relation (3.72) will take the form $f = \frac{\omega}{\pi} \approx 8960\sqrt{n} \approx 10\ \sqrt{n}$, therefore, the plasma frequency of the solar irradiation with electron concentration of $n \sim 10^8$ cm^{-3} will be in the level of ~100 MHz (VHF waves). Conductivity of plasma is expressed through the plasma frequency also, as follows:

$$\sigma = \omega_0^2 \tau = \frac{q^2 n\tau}{m} = \frac{q^2 n\lambda}{mv_e} \tag{3.73}$$

where λ is the mean free path of electrons and v_e is their random heat velocity.

A glow discharge (plasma) in the vacuum chamber strives to occupy the whole volume of the chamber due to the diffusion processes. However, there are the dark gaps between the plasma and walls of the chamber. The plasma loses electroneutrality near the solid surfaces due to the difference in the velocities of electrons and ions. At the same average energy, the electrons' velocity will be more than the ions' velocity, several hundred times. Due to the high escape speed of electrons, they can reach the chamber's wall faster than ions and thus charge them up negatively with respect to the plasma. The main part of the potential difference between the solid and the plasma will be confined to a narrow sheath restricted to several Debye lengths in thickness. It is usually believed that the space charge has a thickness of ~$4\lambda_D$.

3.4 Magnetron sputtering

Sputtering is one of the most effective methods of thin film deposition. It is widely used in the present-day semiconductor, photovoltaic and microelectronics industries. Materials with high melting points such as ceramics and refractory metals, which are hard to deposit by evaporation, are easily deposited using sputtering. Different sputtering methods, from a simple DC glow discharge sputtering to the multi-electrode DC and RF sputtering, enable deposition of any type of material and synthesis of new materials with pre-defined properties. For instance, sputtering can be used to create a new metal material from an intermetallic compound, this new material has the same chemical composition as an intermetallic semiconductor with a zinc-blend or wurtzite-type crystalline structure, and is stable at room temperature and atmospheric pressure. At the same time, each sputtering method has its particular limitation (for example, DC sputtering requires a conductive target only, etc.).

Sputtering is a process in which highly energetic ions strike the surface of a solid target, causing the emission of neutral atoms or molecules from the target surface as a result of momentum transfer. Assuming a hard-sphere elastic-collision model, the exchange of momentum among atoms in and around a collision cascade is a statistical process similar to that encountered in billiards. This process is illustrated in Fig. 3.19.

Upon the bombardment of the target with ions, the following *p* processes may occur:

1. The incident ion is reflected back and gets neutralized in the process.
2. Secondary electrons are emitted due to ion impact.

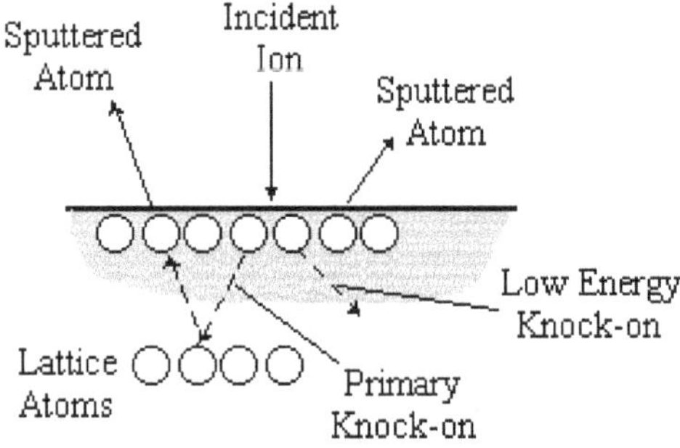

Fig. 3.19 Momentum exchange processes at the target.

3. The incident ion causes ejection of the atom/atoms of the target.
4. The ion itself is buried in the target with/without simultaneous target atom ejection.

Consideration of the momentum transfer between the incident and the target particles reveals this transfer to be at its maximum when the particle masses are equal. The maximum momentum transfer from the incident ions correlates to the maximum target atom sputter rate. The average energy of the atoms sputtered from the surface lie in the range of 10–40 eV. However, the energy of the sputtered particles can reach that of the sputtering ions. Typical sputter yields for metals lie in the range of 0.5–2 atoms/ion.

All sputtering methods can be divided into two main categories. In the first category, the plasma occupies the whole volume of the vacuum chamber and comes in contact with the grown layer and with the walls of the vacuum chamber. In the second category, the plasma is confined to a restricted volume by means of electromagnetic fields or other methods. In this case, independent control of different physical parameters affecting the deposition process becomes especially important.

Each independent physical deposition parameter, such as sputtering voltage, substrate temperature or gas pressure, can be considered as a degree of freedom in the deposition process. Controlability of the required coating properties increases with the number of degrees of freedom. All known sputtering methods are based on a number of variable input parameters that are independent and controllable. For example, the distance between the plasma and the sputtering target may be used as an independent and controllable parameter. This parameter permits to modify the film growth rate independently of all other deposition parameters.

Originally, a physical sputter deposition system with DC diode arrangement, as shown in Fig. 3.17, simply represents two parallel plates powered by a power supply of several kilovolts. The sputtering process here was conducted under a working pressure of several tens to hundreds of mTorr. The negative plate, also known as the cathode or the sputtering target, was bombarded with ions from the plasma volume located between these two plates. The sputtered cathode atoms could then deposit on various surfaces inside the vacuum chamber, forming films. The DC diodes were characterized by slow deposition rates, high voltage and low currents. This method was not suitable for deposition of dielectric and semiconductor films. Replacing the DC supply with an RF supply, usually operated at a frequency of 13.56 MHz, made deposition of dielectric films possible. However, the deposition rate was still low. A further improvement in sputtering techniques was achieved by confining plasma electrons by means of a magnetic field. This technique is known as magnetron sputtering. Higher plasma density reduces the discharge impedance and results in much higher current and lower voltage discharge.

The electron confinement in a magnetron system is due to the presence of orthogonal electrical ($\boldsymbol{E}$) and magnetic ($\boldsymbol{B}$) fields near the cathode surface. These fields result in a classical $\boldsymbol{E} \times \boldsymbol{B}$ drift for electrons (the Hall effect), which gives rise to a sequence of cycloidal hopping steps parallel to the cathode surface. As a result, the secondary electrons, which are emitted from the cathode due to ion bombardment, are confined to the nearest vicinity of the cathode. In a magnetron, the electric field is always oriented normally to the cathode surface. The transverse magnetic field is configured in such a way that the $\boldsymbol{E} \times \boldsymbol{B}$ drift paths form closed loops, which trap the drifting electrons and force them to circulate many times around the cathode surface. As a consequence, the ions sputter only a limited part of the cathode and create heterogeneous etching ring zones on its surface. Figure 3.20 shows a side view of the simplest arrangement of the magnetron system and the experimentally obtained structure of the magnetic field near the target surface.

As shown in Fig. 3.20(a), the simplest magnetron system consists of the permanent magnet, a magnetic system made of the soft steel, the sputtering target (cathode) and an anode situated at a short distance from the cathode. Figure 3.20(b) represents the magnetic field distribution near the cathode reconstructed using the iron filings. The magnetron represents an axis-symmetric construction. Consequently, magnetic and electric fields in this design are also axis-symmetric.

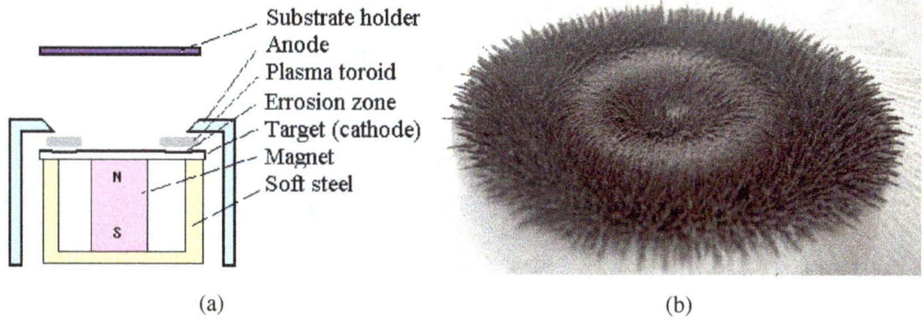

Fig. 3.20 The simplest magnetron sputtering system: (a) a schematic side view of the magnetron sputtering source, (b) the magnetic field distribution near the cathode.

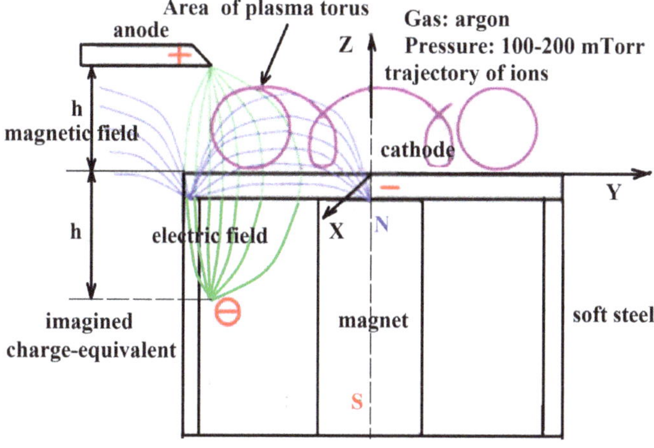

Fig. 3.21 A theoretical scheme of the planar circular magnetron.

Figure 3.21 represents a theoretical scheme of the planar circular magnetron.

In this system, the cathode (target) has a high negative potential with respect to the grounded anode. Two different forces created by the electric field between the anode and cathode and the magnetic field from a permanent magnet located inside the cathode affect the behavior of charge carriers.

Every charge is a source of the electric field, through which it exerts force on any other charge in inverse square proportion to the distance between them. An external electric field will apply forces to any charge that enters it. Charges moving relative to magnetic fields also experience forces. The Lorentz force, F_L, conveniently combines the electric and magnetic effects on a charge:

$$\boldsymbol{F}_L = m\frac{d\boldsymbol{v}}{dt} = q\left(\boldsymbol{E} + \boldsymbol{v} \times \boldsymbol{B}\right) \tag{3.74}$$

Here, q is the elementary charge in Coulombs; $\boldsymbol{E}$ is the electric field in Volts per meter; v is the velocity in meter per second; × implies the vector product and $\boldsymbol{B}$ is the magnetic flux density in Tesla. In the case of constant electric field $E = E_0$ without a magnetic field ($B = 0$), a particle moves with a constant acceleration along E_0:

$$z(t) = z_0 + v_0 t + \frac{qE_0}{2m}t^2. \tag{3.75}$$

where z_0 and v_0 are the particle position and velocity at $t = 0$. Evidently, the motion direction is defined by the type of charged carriers, positive or negative.

For the movement of charged carriers in the magnetic field ($B = B_0$) without an electric field ($E = 0$), our equations will take the following form:

$$\begin{cases} m\dfrac{dv_x}{dt} = q\left(v \times B\right)_x = qv_y B \\ m\dfrac{dv_y}{dt} = -qv_x B \\ m\dfrac{dv_z}{dt} = 0 \end{cases} \tag{3.76}$$

Combination of the first two equations gives a motion equation in the direction y:

$$\frac{d^2 v_y}{dt^2} = -\left(\frac{qB_0}{m}\right)^2 v_y \tag{3.77}$$

This equation describes the circular motion of a charged particle with gyration (cyclotron) frequency equal to

$$\omega_c = \frac{qB_0}{m} \tag{3.78}$$

Solving Eq. (3.77) using (3.78) to obtain v_x, we have

$$\begin{cases} v_y = v_{\perp 0}\cos(\omega_c t + \varphi_0) \\ v_x = v_{\perp 0}\sin(\omega_c t + \varphi_0) \\ v_z = v_{z0} \end{cases} \tag{3.79}$$

where $\nu_{\perp 0}$ is the speed component perpendicular to B_0 (y-axis), and φ_0 is an arbitrary phase. Integrating yields the particle position

$$\begin{cases} y = r_c\sin(\omega_c + \varphi_0) + (y_0 - r_c\sin\varphi_0) \\ x = r_c\cos(\omega_c + \varphi_0) + (x_0 - r_c\cos\varphi_0) \\ z = z_0 + v_{z0}t \end{cases} \tag{3.80}$$

where

$$r_c = \frac{v_{\perp 0}}{|\omega_c|} \tag{3.81}$$

Equation (3.80) show that the particle moves in a circular orbit perpendicular to B, with frequency ω_c and radius r_c, around a guiding center with coordinates $(x_0, z_0, y_0 + v_{y0}t)$, which moves uniformly along the magnetic field force lines. Positive ions gyrate around the magnetic field lines according to the left-hand rule, and electrons gyrate according to the right-hand rule. This motion is shown in Fig. 3.22.

In the case of two constant fields, electric $E = E_0$ and magnetic $B = B_0$, the negative particles will obtain drift in the negative direction

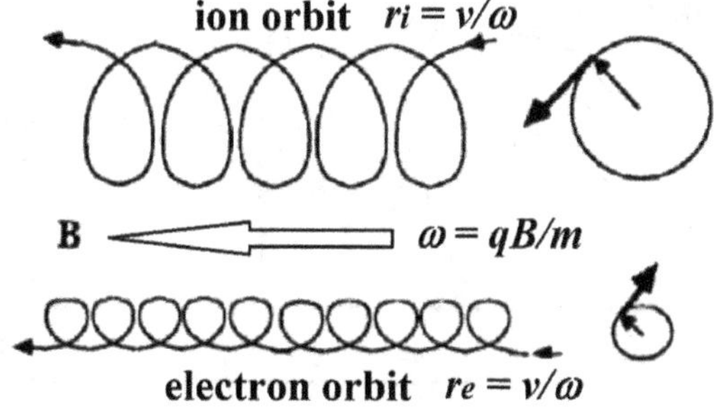

Fig. 3.22 Movement of charge carriers in the magnetic field.

of the axis X according to the right-hand rule. Now, the components of Eq. (3.74) will look as follows:

$$\begin{cases} m\dfrac{dv_x}{dt} = q\left(v\times B\right)_x = qv_y B \\ m\dfrac{dv_y}{dt} = q\left[E+\left(v\times B\right)\right]_y = qE - qv_x B \\ m\dfrac{dv_z}{dt} = 0 \end{cases} \tag{3.82}$$

In this case, electrons experience constant force resulting in the cycloidal motion with ultimate movement to *x* direction, i.e., motion perpendicular to the magnetic and electric field force lines. Figure 3.23 illustrates the behavior of charged carriers in the magnetron conditions.

A magnetron construction contains only two electrodes, anode and cathode. However, the main difference between the diode sputtering system (see Fig. 3.17) and the magnetron sputtering system shown in Fig. 3.23 is in the confinement of electrons and ions in the trap created by crossing of orthogonal electric and magnetic fields.

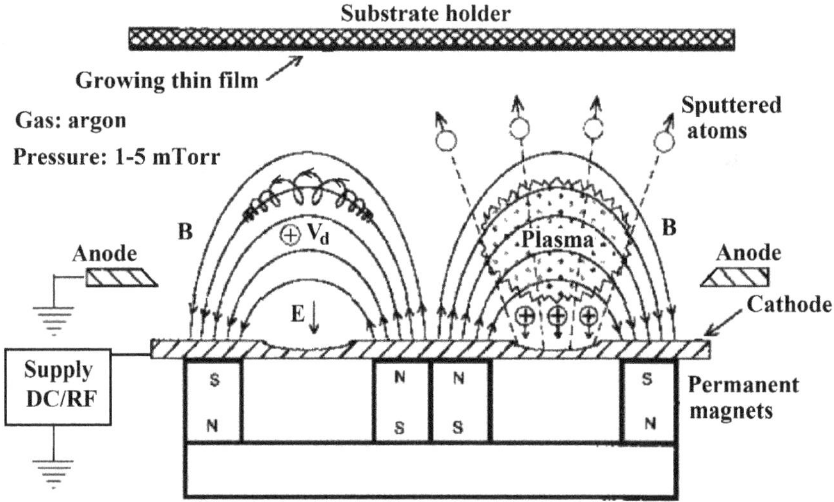

Fig. 3.23 The plasma behavior in the planar circular magnetron conditions.

Due to the increase in the mean free path of electrons in this trap, they can produce a large number of ions because of the many collisions of electrons with atoms of a neutral gas. This results in possibility to decrease the gas pressure in the vacuum chamber and to work in the region of the reduced product *pd* in Eq. (3.44) of the Paschen law. Wherein, the plasma discharge becomes an abnormal plasma discharge, when a plasma current is directly proportional to the applied voltage. Due to the reduced vacuum pressure, the number of collisions of the sputtered atoms with the neutral gas atoms also decreases and the sputtered atoms come to the substrate with elevated energy that affects the growing thin films. The growing thin films have increased adhesion and density. The magnetron sputtering significantly increases the deposition rate compared with the diode sputtering systems. Also, the magnetron sputtering method makes it possible to grow thin films of all types: metallic and alloy thin films as well as semiconductors and dielectric coatings using DC or RF sputtering supplies.

The main parameters of plasma are the temperature of electrons and the concentration of positive ions which may be used for sputtering. One of the simple and effective methods for studying the plasma parameters is the Langmuir probe method. A metal probe inserted in a discharge gap, and biased positively or negatively to draw the electron or the ion current, is one of the most efficient tools for plasma diagnostics. This probe, called the Langmuir probe, together with its typical voltage–current characteristic is shown in Fig. 3.24. In this picture, V_B represents the bias potential on the probe, Φ_f is known as the floating potential and Φ_p is the plasma potential.

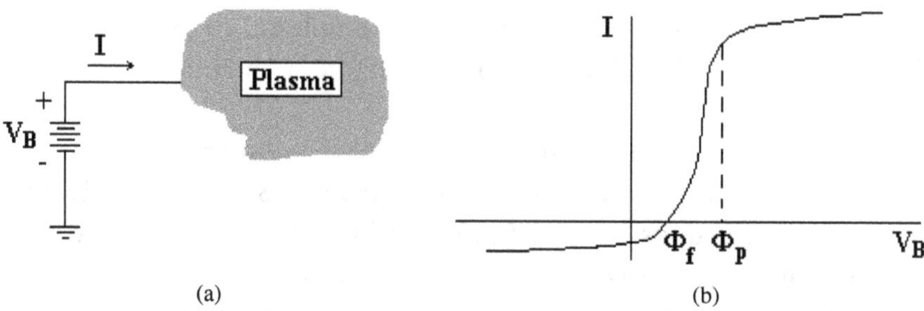

Fig. 3.24 Definition of voltage and current for a Langmuir probe (a) and a typical *I–V* characteristic for a Langmuir probe (b).

At the probe voltage $V_B = \Phi_p$, the probe is at the same potential as the plasma and mainly draws current from the more mobile electrons. This current is designated as positive current flowing from the probe into the plasma. For V_B exceeding this value, the probe current tends to saturate at the electron saturation current. The saturation current is defined by the probe geometry. At $V_B < \Phi_p$, electrons are repelled in accordance with the Boltzmann relationship, until at Φ_f the probe is sufficiently negative with respect to the plasma, so that the electron and ion currents are equal, and $i = 0$. For $V_B < \Phi_f$, the current is mainly ion current (negative with respect to the plasma) and tends to saturate at the ion saturation current, which may also vary with the applied voltage and due to changes in the probe effective collection area. The value of the ion saturation current is much lower than that of the electron saturation current due to a much greater ion mass. The electron temperature may be estimated from the probe current graph at $\Phi_f < Vb < \Phi_p$ region in Fig. 3.24 using the Boltzmann distribution for electrons.

The measuring scheme and the probe construction are dictated by the experimental conditions and the plasma parameters. The type of model to be best used for estimation of the plasma characteristics is defined by the interrelation between three basic parameters: r_p (the probe radius), λ_D (Debye radius, see formula (3.67) and λ (the mean free path). As mentioned above, the probe is surrounded by a layer of volume charge. This layer's thickness is very important for the measurement. One can describe three typical cases:

1. $\lambda >> r_p >> \lambda_D$, thin layer of volume charge;
2. $\lambda >> \lambda_D >> r_p$, collisionless thick layer of volume charge;
3. $\lambda_D >> \lambda >> r_p$, thick layer of volume charge with collisions.

Figure 3.25 presents a side view of the Langmuir probe used in studying the magnetron system shown in Figs. 3.20 and 3.21. The probe made of tungsten wire 0.25 mm in diameter was placed in a ceramic tube. The active part of the probe was 3.4 mm long.

If the pressure in the vacuum chamber does not exceed the level of $1 \cdot 10^{-3}$ Torr, the mean free path according to formula (2.11) will be equal to $\lambda = 5$ cm. Weakly ionized plasma contains approximately

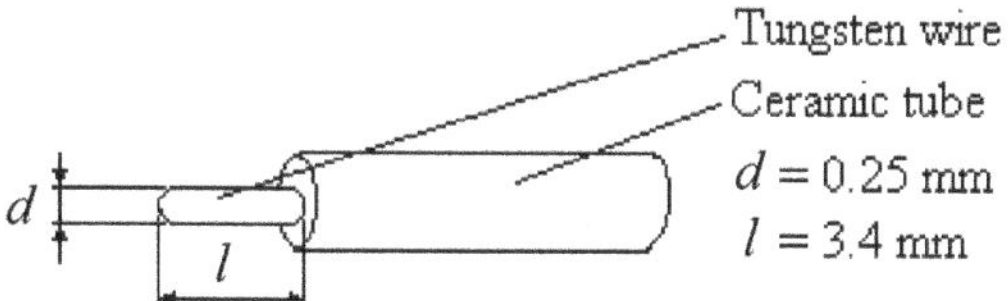

Fig. 3.25 A side view of the applied Langmuir probe.

0.01 % of ionized particles in the volume unit. The gas molecule concentration can be obtained from the known formula (2.1) $p = n_g kT$, where p is the pressure in the vacuum chamber, n_g – molecule concentration, and T – the gas temperature. Therefore, the electron concentration in the argon atmosphere with the pressure of $1 \cdot 10^{-3}$ Torr is approximately $3.56 \cdot 10^9$ cm^{-3}. If we assume that the electron temperature of the weakly ionized plasma is ~1–10 eV, the Debye radius will be equal to approximately $\lambda_D = 0.125–0.395$ mm (from formula (3.67)). Comparison of the parameters r_p, λ_D and λ shows that they satisfy the second case: $\lambda >> \lambda_D >> r_p$: collisionless thick layer of the volume charge. In this case, we can estimate the electron temperature in the plasma from the experimental I–V characteristics of the Langmuir probe, using the following known formula:

$$T_e = \frac{e}{k}\left(\frac{d \ln J_e}{dV_B}\right)^{-1} \tag{3.83}$$

In such a way, the electron temperature is equal to the slope angle of the Langmuir probe characteristic in the semi-logarithmic scale. This "ideal" characteristic ignores the "perturbation" processes such as bombardment of the probe by high-energy electrons, emission of secondary electrons from the probe, and etching away of the probe. The collisionless condition allows us to use the Bohm approximation to estimate the ion concentration in the plasma:

$$j_{is} \approx en_i v_B = en_i \sqrt{\frac{kT}{m_i}} \tag{3.84}$$

where j_{is} is the probe current saturation density and v_B is the Bohm velocity.

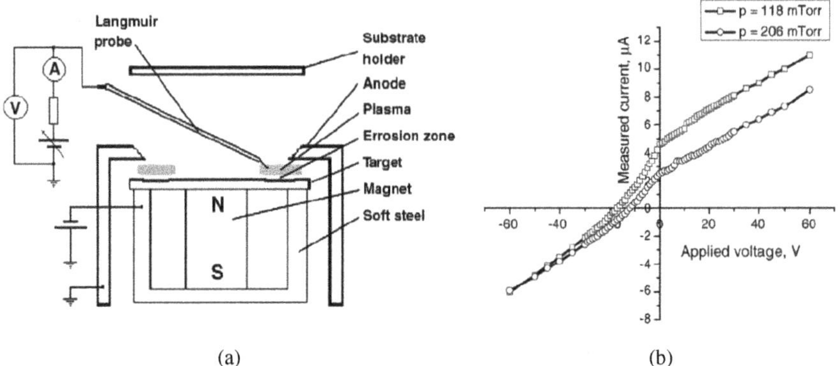

Fig. 3.26 Schematic view of a magnetron sputtering system with a Langmuir probe (a), Langmuir probe characteristics measured with this system (b).

Method of the Langmuir probe measurement was successfully applied to characterization of various configurations of plasma discharges. For example, Fig. 3.26 represents the schematic view of the simple magnetron sputtering system (see Fig. 3.20) and *I–V* Langmuir characteristics measured in this system.

As shown in Fig. 3.26(b), the *I–V* characteristics are significantly influenced by the pressure in the vacuum chamber. In cases of these relatively high gas pressures, a little decreasing of the pressure leads to the growth of the measured electron current. This is explained by decreasing of collisions of electrons with the pressure decreasing and consequent current increase. The *I–V* curves measured using the Langmuir probe enable precisely defining the floating potential and the plasma potential of the gas discharge. On the basis of these measurements, we can calculate the temperature of electrons in plasma and the ion spatial distribution. Calculations show that the maximum electron temperature (~6.5 eV) and the maximum ion concentration ($\sim 9.5 \cdot 10^{12}$ cm^{-3}) in given system configuration relate to the maximum gradient of the magnetic field.

Figure 3.27 represents the recent magnetron sputtering source, a magnetic field distribution reconstructed using the Ni powder and the side view of the magnetron discharge in argon.

The present magnetron sputtering sources contain the modular high-strength neodymium magnets providing the required magnetic

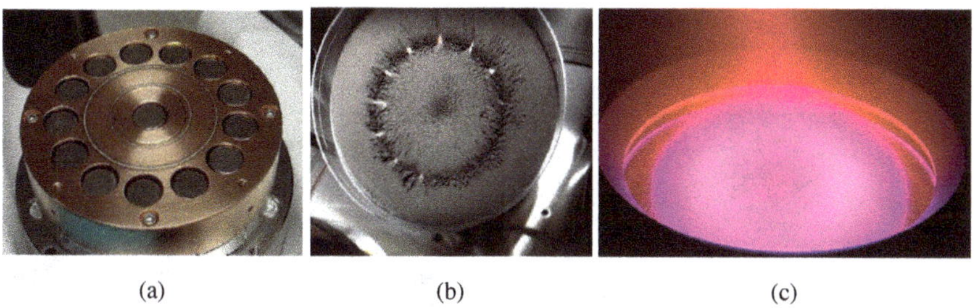

(a) (b) (c)

Fig. 3.27 A present magnetron sputtering source (a), a magnetic field distribution reconstructed using the Ni powder (b), a side view on the magnetron discharge in argon (c).

field configuration. This magnetic field enables reduction of the argon working pressure by two orders of magnitude compared with the simple magnetron system. Now, the magnetron self-sustained discharge can work stably at pressure of 1–5 mTorr. Mean free path of gas molecules reach 5 cm at this pressure and mass transfer under sputtering goes into ballistic type from the diffusion type. Thus, sputtered atoms at this pressure can reach the substrate without thermalization and with high energy, enough to create the strong bonds with the substrate surface. Construction of the magnetron sputtering sources permit to provide co-sputtering processes as shown in Fig. 3.28.

A vacuum chamber equipped with two or more independent sputtering sources enables growth of complex thin films with predicted properties. One of the sources supplied with the DC sputtering can produce metal films and the second one, equipped with the RF sputtering, can produce oxide, nitride or complex semiconductor materials. The simultaneous sputtering process enables growth of the semiconductor films with predicted impurity concentration. For example, by this method may be grown such thin films as ZnO alloyed with Y or ZnS alloyed with Mo, Si alloyed with Ge, etc. Each sputtering source, in the magnetron system, may be controlled independently of others. The sputtering power, distance between the magnetron and the substrate holder, gas environment in the vicinity of the target, all these process parameters may be controlled independently. The common parameters in the process are the substrate

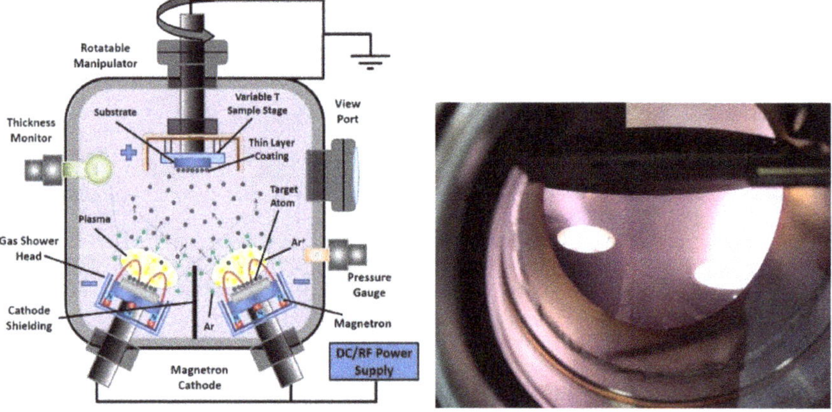

Fig. 3.28 Co-sputtering of two different materials from two magnetron sources: (a) a schematic view of the vacuum chamber and (b) the simultaneous sputtering from two sources.

temperature and the substrate rotation rate. The heater arranged with the substrate holder enables heating the substrate up to 700–800°C, which permits to control the crystallinity of the growing thin film.

Sputtering is caused by striking the target surface by incident ions. Thus, the sputtering rate, R_s, will be proportional to the number of incident ions and the density of the sputtered target.

$$R_s = S\frac{j_i}{q}\frac{1}{\gamma_t} \tag{3.85}$$

where j_i is the ion current density, γ_t is the sputtering target density and S is the proportionality coefficient called the sputtering yield. Sputtering yield may be defined as the removal rate of the target surface atoms due to ion bombardment. The sputtering yield may be described by the following empirical relation:

$$S = \frac{\text{mean number of emitted atoms}}{\text{incident ion}} \Rightarrow S \propto \frac{M_{\text{ion}} M_t}{\left(M_{\text{ion}} + M_t\right)^2}\frac{\ln E}{E}\frac{1}{cos\theta} \tag{3.86}$$

where M_{ion} is the ion mass, M_t is the target material atomic mass, E is the mean energy of ions and θ is the incident angle of the coming

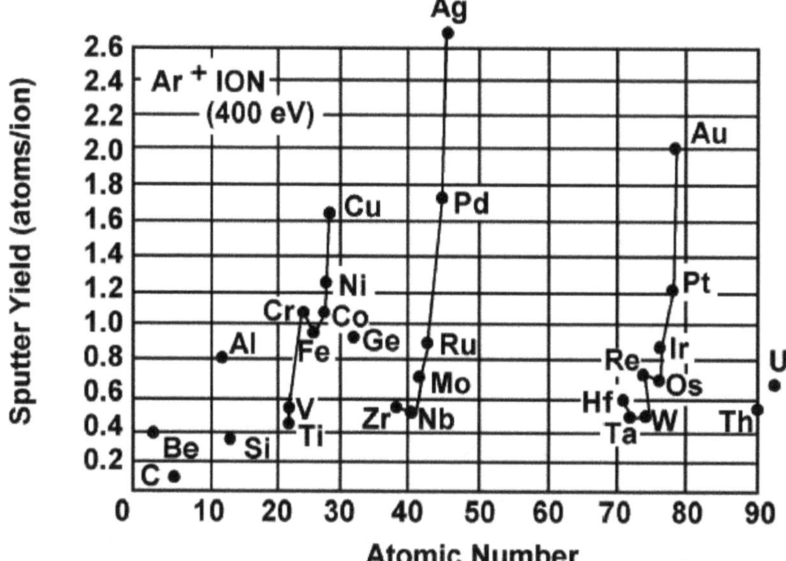

Fig. 3.29 Sputtering yield distribution dependent on the atomic number.

ions. Thus, the sputtering yield depends on the composition and the crystal structure of the sputtering target, the energy of incident ions and on the incident angles of coming particles. Energy of incident ions is defined by voltage applied to the cathode relative to the ground. The plasma potential usually does not prevail with several tens of Volts, but supply applied to the cathode is in the interval of 400–5000 V. This makes it possible to obtain neutral sputtered atoms with an average energy of 15–20 eV, which is sufficient for thin films growth. Figure 3.29 presents the periodic diagram of the sputtering yield distribution dependent on the atomic number of the sputtered material under incident argon ions with average energy of 400 eV. This data may be used as a reference material when developing suitable deposition technology.

Energy of sputtering ions and incident angle also significantly influence the sputtering yield as shown in Fig. 3.30.

All these deposition parameters influence the properties of thin films deposited by sputtering and they must be taken into account when developing the specific technology of the thin film growth. For

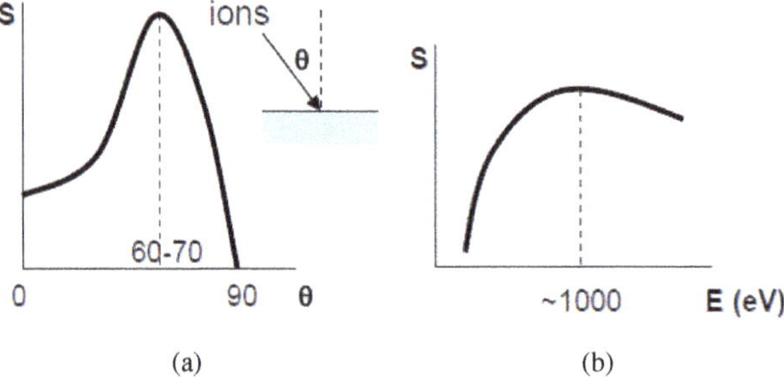

Fig. 3.30 Sputtering yield dependence on the incident angle (a) and the energy of sputtering ions.

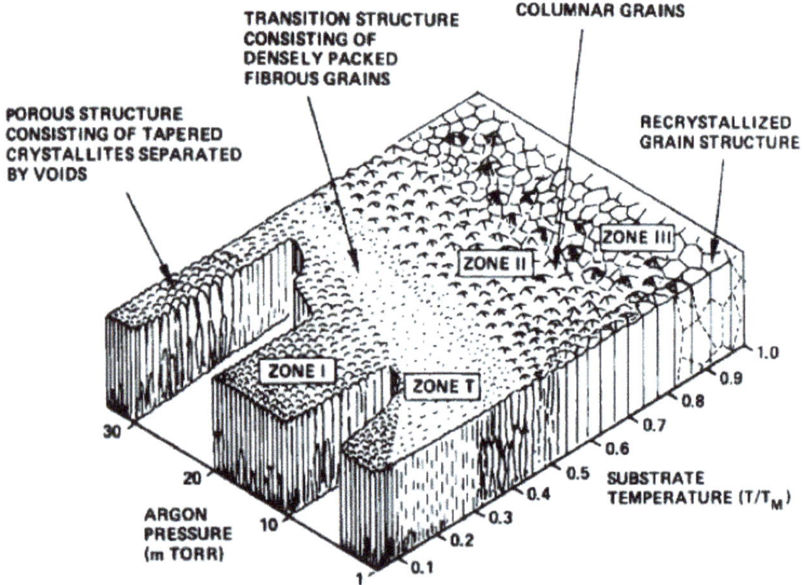

Fig. 3.31 The structure of a deposited metal film dependent on the gas pressure and the substrate temperature.

example, the microstructure of thin films deposited by sputtering significantly depends on the level of gas pressure in the vacuum chamber and the substrate temperature. The basic model for the structure of metal thins films deposited by sputtering was developed by Movchan and Demchishin and afterwards modified by Thornton. Figure 3.31 illustrates the modified model of thin films growth.

The substrate temperature in the presented model is shown as homologous temperature, that is the temperature normalized to the melting point. One can see from this model that the substrate temperature growth leads to the crystallites' increasing. The presented model shows that using the gas pressure in the vacuum chamber and a temperature of the substrate we can control the crystallinity of the growing films. Subsequent experiments have shown that this model is right also for non-metal thin films, for example ZnO.

Chapter 4

Experiment planning, optimization and data presentation

4.1 Introduction

Growth of a thin film is a complex process influenced by many independent parameters. These independent parameters or variables act together on the process and fully define the properties of a grown film. Sometimes, these parameters are called the degrees of freedom. Thus, a number of independent influencing parameters will represent the level of freedom for the system or the process. For example, for the vacuum evaporation process, the value of the power supply, the wall and the substrate temperatures, the value of residual, total and partial pressures of used gases, the vacuum chamber dimensions, the process duration — all these parameters are acting factors, although some of them are independent and some, dependent. Such a system that has several influencing factors is called a multi-parameter system. Evidently, some of the variables act on the properties of the grown thin films more than others. To evaluate the level of influence of various variables on the film's properties, different models of the process may be built. These models may be purely theoretical or empirical, that is based on experiments. Properties of the grown thin films may be designated as the response function or output function. Model of the deposition process will relate values of the response function with input variables. Thus, a model of the process may be considered a transfer function. Experiment design allows us to find a mathematical representation of the desired model. Optimization of such a model makes it possible to obtain the most adequate model with minimal experimental trials.

There are many different methods used to empirically study the relationships between one or more of the measured response functions, on the one hand, and a number of input parameters, on the other hand. All these methods enable us not only to build the required model, but also optimize it and decrease the number of experimental trials. Here, we shortly consider only three methods applied for this goal. These are the multi-simplex method, the Taguchi method and the response surfaces method (RSM).

The simplex method is based on the initial design of $k+1$ trials, where k is the number of variables. A $k+1$ geometric figure in a k-dimensional space is called a simplex. With two variables, the first simplex design is based on three trials. This number of trials is also the minimum required for defining the direction of the process improvement. After the initial trials, the simplex process is sequential, with the addition and evaluation of one new trial at a time. The simplex searches systematically for the best levels of the control variables. Therefore, it is a timesaving and economical way to start an optimization process.

The Taguchi method is based on the design of experiments to provide near optimal quality characteristics for a specific objective. The Taguchi method includes the integration of statistical design of experiments into a powerful engineering process. The goal is to optimize an arbitrary objective function and to reduce the sensitivity of engineering designs to uncontrollable parameters or noise. The objective function used and maximized in the present case is the signal-to-noise ratio. This method involves three steps of the optimization process:

1. System design (development of a system designed to function under nominal conditions, with a given initial set of input parameters);

2. Parameter design (selection of optimal levels for the controllable system parameters);
3. Tolerance design (a narrower tolerance range must be specified for those design factors whose variations impart a large negative influence on the output function).

The Taguchi method can significantly reduce the number of experiments needed to find the response function, by improving the efficiency of generating information needed for the system design.

The response surface method is an optimization approach, which uses mathematical and statistical techniques to search for the best combination of the process variables. The response surface represents the domain of all feasible values of the response function. RSM is the best method for an empirical study of the relationships between one or more of the measured response functions such as resistivity, transparency, density, on the one hand, and various input parameters, on the other hand. RSM enables the process optimization with a minimum of trial and errors. If the process model is presented as a mathematical function, one can say that the response surface is a trace of the response function (the main parameter) in the multi-parameter space. Then it can be written in the form of $F = f(X_1, X_2, .. X_m)$, where X_1, X_2, and X_m are the independently controlled measured input variables of the growth process and the response surface represents the transfer function of input independent factors. Figure 4.1 illustrates a graphical representation of the response surface method. The black box here represents the response function or the transfer function. Each input factor may take different values inside the suitable domain

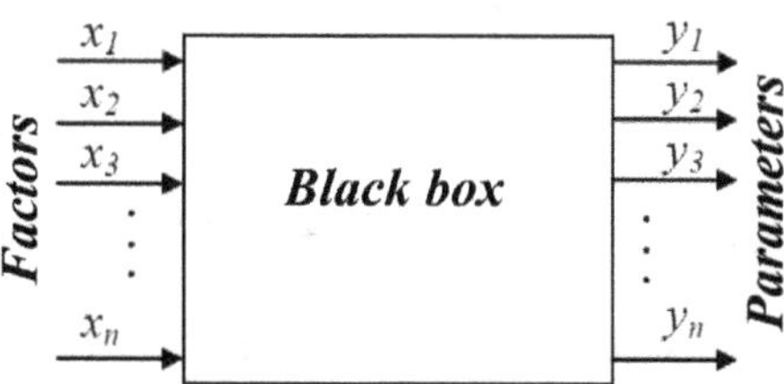

Fig. 4.1 Presentation of the surface response method in the form of the "black box".

of definition. These domains of definition can be continuous and discontinuous, limited and unlimited. Factors are independent and controllable quantities. This representation of the response function allows us to build a process model without information about the internal structure of the black box, using only experimental data.

In the narrow interval of variations of the input variables (factors), the response function (parameter) may be considered as an analytical function. The approximation process may begin as a polynomial model with linear independent coefficients. Therefore, it may be expanded in a Taylor series over the chosen area in the following form:

$$f(x)=\sum_{n=0}^{\infty}\frac{f^{(n)}(x_0)}{n!}(x-x_0)^n \tag{4.1}$$

By choosing the first three terms of the above expansion, a second order approximation is obtained as follows:

$$f(x)=f(x_0)+\frac{df(x_0)}{dx}(x-x_0)+\frac{1}{2}\frac{d^2f(x_0)}{dx^2}(x-x_0)^2 \tag{4.2}$$

This expression has the form of a parabolic function with linear coefficients:

$$f(x) = a + bx + cx^2 \tag{4.3}$$

For a multi-parameter process of ***m*** levels, i.e., with m independently controlled measured input variables, the expression is transformed to a model of the following type, without the high-order (in our case, third-order) interaction effects:

$$\begin{aligned} F= c_0 + c_1X_1 + c_2X_2 + c_3X_3 + \cdots + c_mX_{1m} + c_{m+1}X_1X_2 + c_{m+2}X_1X_3 \\ + \cdots + c_kX_{m-1}X_m + c_{k+1}X_1^2 + c_{k+2}X_2^2 + \cdots + c_{k+m}X_m^2 \end{aligned} \tag{4.4}$$

where F is the response function, c_i are the model coefficients and X_i are the processing parameters. The response function exists in the space called factor space, and the number of independent factors

Table 4.1 Standard form for model coefficient estimation.

N	X_1^*	X_2^*	X_3^*	...	X_m^*	F
1	+1	+1	+1	...	+1	
2	+1	+1	+1	...	–1	
...	...	...	...	...	...	
2^m	–1	–1	–1	...	–1	

Note: F is the response function and X_i^* are the normalized input process factors.

determines the dimension of the factor space. A model presented in the form of Eq. (4.4) is called the regression model and its coefficients, c_i, are regression coefficients. A standard form for a designed series of experiments for evaluating the normalized process factors is shown in Table 4.1. This form is also known as a plan matrix or the orthogonal plan matrix. The input factors are normalized using the following equation:

$$X_i^* = 2(X_i - X_0)/\Delta X \qquad (4.5)$$

where X_0 is the midpoint of the variable range, and ΔX is the variation range.

Obviously, in order to create a complete multi-factor space from the m variable system, a minimum of 2^m separate experiments are required; the constants can then be calculated from the 2^m resulting equations. In addition, a few more experiments in the system center ($X_i^* = 0$) are required in order to estimate the error and reliability of the calculated result. However, from a practical point of view, there is no need to run all of these experiments. It is possible to complete only a few experiments and then deduce the missing coefficients using the mathematical procedure known as "estimation on partial sections in multi-parameter space". According to this procedure, each one of the experiments is actually a partial section of the chosen multi-parameter space. Consequently, the coefficients of the approximation model must be related for all partial sections. Thus, the missing coefficients can be calculated by varying one parameter in the following quadratic model:

$$\begin{cases} c_1 + c_2 x + c_3 x_2 = F_k(x = x_k) \\ c_1 + c_2 x + c_3 x_2 = F_{k+n}(x = x_{k+n}) \\ c_1 + c_2 x + c_3 x_2 = F_{k+m}(x = x_{k+m}) \end{cases} \tag{4.6}$$

where c_i is the missing coefficient, x is the process parameter $(X_1, X_2, ..X_N)$, F is the response function and m, n are indices. The rest of the missing coefficients describing the required model are then obtained by solving the coefficient matrix with the Kramer method. In the case of a large deviation between the measured data and the model, the approximation order, i.e., the order of approximation equation, should be increased.

4.2 Mathematical modeling of the thin film deposition process

An experiment is the main and most perfect method of cognition. It can be active or passive. The implementation of a passive experiment does not depend on the experimenter, and he has to be content with only the role of an observer. The main type of experiment is the active one, carried out under controlled conditions. Usually, it is impossible to foresee all the factors influencing the investigated parameters of the object. Thus, in complex systems that depend on many factors, some impacts cannot be controlled or managed. The effect of these factors is considered as white noise superimposed on the true results of the experiment. To separate the factors of interest to the experimenter from the background noise, special methods are used, called the experiment randomization. Randomized experiments are experiments that provide the most reliable and solid statistical estimates of the effect of operating factors.

Evidently, not all input factors equally influence the studied output parameter. So, before beginning to design the plan for an experiment, it is desired to study all existing information about the technological process. This will help to design a more reliable plan and exclude from the plan the combinations of input factors which slightly affect the output parameter. Sometimes a design workflow contains multiple output parameters that can conflict with one another. For example, to create a transparent conductive coating, it is necessary to obtain two output parameters with optimal properties: a maximum transmittance in a given optical range and a maximum possible conductivity. These two parameters are contradictory. To be transparent in the visual range, the transparent conductive layer should represent the wide bandgap semiconductor doped by the suitable impurities. Increasing the impurity concentration, and as a result increasing the conductivity in the material, leads to decrease in transparency. Thus, one of the output parameters will be chosen as the response function and the second one will play the role of a limitation, influencing the choice of the possible combination of active input factors.

The first and very important stage in developing an experiment model is the experiment statement. This includes formulating the main goal of the experiment. This stage also includes the selection of independent input factors and the determination of the output function and limitations.

Now, to create the full factor space or to design the full factor experiments and to obtain a response function surface, all possible combinations of the active input factors need to be realized. In this process, for each factor, two levels are selected — upper and lower — at which the factor varies. Half of the difference between the upper and lower levels is called the variation interval, which should be more than the measurement error. As it was mentioned above, the number of trials in the case of m factors will be equal to 2^m. The designed orthogonal plan enables to build the surface function in the m-dimension factor space and offers the possibility to estimate influence of each factor on the built model.

The built process model must be precise. To estimate the precision of the model at each point of the factor space, a number K of tests should be performed. These tests should be done by random selection of the points. This enables us to evaluate the reproducibility and repeatability of the process, calculate the average values of the measured parameters and estimate the adequacy of the built model using for example the least squares method. Average value of the response function in the *i*th point will be calculated by the following formula:

$$\overline{f_i} = \frac{1}{K}\sum_{j=1}^{K} f_{ij} \tag{4.7}$$

The test is considered reproducible and repeatable if the dispersion D_{fi} of the output parameter f_i is homogeneous at each point of the factor space. The standard deviation S_{fi} of the dispersion D_{fi} is determined for each point of the factor space by the following formula:

$$S_{fi}^2 = \frac{1}{1-K}\sum_{j=1}^{K}\left(f_{ij} - \overline{f_i}\right)^2 \tag{4.8}$$

The hypothesis of homogeneity (equality) of dispersions is checked using the Cochran test. The calculated value of this criterion, *Ci*, is determined by the following formula:

$$C_i = \frac{\max S_{fi}^2}{\sum_{j=1}^{K} S_{fi}^2} \tag{4.9}$$

And the critical value of the criterion C_{cr} is found from the Cochran distribution table by the number of degrees of freedom of the numerator $(K - 1)$, the denominator K and the level of significance q (see Table 4.2). If $C_i < C_{cr}$, the hypothesis of homogeneity of dispersions is accepted, otherwise — rejected, and then the experiment must be repeated, changing the conditions of its implementation. Usually, in technical problems the value $q = 0.01 \div 0.05$ is chosen, which corresponds to the $(1 \div 5)\%$ significance level.

The number of tests in the complete factorial experiment 2^m grows rapidly with an increase in the number of input factors m, and for large m, this type of experiment planning turns out to be practical.

Table 4.2 The table of C_i-distribution.

n_2 | n_1	1	2	3	4	5	6	7	8	10
2	0.998	0.975	0.94	0.91	0.86	0.85	0.83	0.82	0.79
	0.999	0.995	0.99	0.98	0.96	0.94	0.92	0.90	0.88
3	0.966	0.87	0.80	0.75	0.71	0.68	0.65	0.63	0.60
	0.991	0.94	0.88	0.83	0.79	0.76	0.71	0.69	0.67
4	0.91	0.77	0.68	0.63	0.59	0.56	0.54	0.52	0.49
	0.97	0.96	0.78	0.72	0.68	0.64	0.61	0.58	0.55
5	0.84	0.68	0.60	0.54	0.51	0.48	0.46	0.44	0.41
	0.93	0.79	0.70	0.63	0.59	0.55	0.52	0.50	0.47
6	0.78	0.62	0.53	0.48	0.45	0.42	0.40	0.38	0.36
	0.88	0.72	0.63	0.56	0.52	0.49	0.46	0.44	0.40
7	0.73	0.56	0.48	0.43	0.39	0.37	0.36	0.34	0.32
	0.84	0.66	0.57	0.51	0.47	0.44	0.46	0.39	0.36
8	0.68	0.52	0.44	0.39	0.36	0.34	0.32	0.30	0.28
	0.79	0.62	0.52	0.46	0.43	0.39	0.37	0.35	0.33
9	0.64	0.48	0.40	0.36	0.33	0.31	0.29	0.28	0.26
	0.75	0.57	0.48	0.43	0.39	0.36	0.34	0.32	0.30
10	0.60	0.45	0.37	0.33	0.30	0.28	0.27	0.25	0.24
	0.72	0.54	0.45	0.39	0.36	0.33	0.31	0.30	0.27
12	0.54	0.39	0.33	0.29	0.26	0.24	0.23	0.22	0.21
	0.65	0.48	0.39	0.34	0.31	0.29	0.27	0.25	0.23
15	0.47	0.33	0.28	0.24	0.22	0.20	0.19	0.18	0.17
	0.57	0.41	0.33	0.29	0.26	0.24	0.22	0.21	0.19
20	0.39	0.27	0.22	0.19	0.17	0.16	0.15	0.14	0.130
	0.48	0.33	0.27	0.23	0.20	0.19	0.17	0.16	0.150
24	0.34	0.24	0.19	0.17	0.15	0.14	0.13	0.116	0.111
	0.42	0.29	0.23	0.20	0.18	0.16	0.15	0.142	0.133
30	0.29	0.20	0.16	0.14	0.12	0.11	0.106	0.100	0.092
	0.36	0.24	0.19	0.16	0.15	0.13	0.123	0.116	0.105
40	0.24	0.16	0.13	0.11	0.10	0.089	0.083	0.078	0.071
	0.29	0.19	0.15	0.13	0.11	0.103	0.095	0.090	0.082

(*Continued*)

Table 4.2 (*Continued*)

n_2 | n_1	1	2	3	4	5	6	7	8	10
60	0.17	0.11	0.09	0.08	0.068	0.062	0.058	0.055	0.050
	0.22	0.14	0.11	0.09	0.080	0.072	0.067	0.063	0.057
120	0.10	0.06	0.05	0.042	0.037	0.034	0.030	0.029	0.027
	0.027	0.08	0.06	0.049	0.043	0.039	0.036	0.033	0.030

Note: C_i is a random variable, distributed according to Cochran's law with suitable degrees of freedom.

To reduce the number of experiment trials, a certain part of them can be selected from the set of points of the factor space. This plan contains a suitable number of trials and represents a fractional factorial design. If the number of factors is $m > 4$, the effects of high-order interactions become statistically insignificant, i.e., the influence of the multiplied factors $X_1X_2X_3\ldots X_m$ on the response function f is mutually compensated. Experimental practice allows us to assume *a priori* that in the regression equation (model) with a large number of factors, the coefficients of high orders of interaction are equal to zero. Therefore, with a large number of factors, it is possible to design such experimental plans that enable one to determine the linear effects of factors or the effect of their paired mutual interactions. If the model seems inadequate, the order of interaction may be increased.

Let us consider the development of a process for the deposition of transparent conductive thin In_2O_3 films by sputtering. Indium oxide thin films possess the electronic properties of n-type semiconductors with a band gap in the range of 3.5–4 eV. Oxygen vacancies in the films provide free electrons by acting as doubly charged donors. Since the indium oxide films generally suffer from oxygen deficiency, the electron gas in the conduction band is degenerate. These transparent films are often alloyed with tin oxide (5–10%); such films are known as indium tin oxide (ITO) films. These films have a higher free charge concentration, which increases the electrical conductivity. Films without the tin admixture have higher resistivity and lower transparency than ITO. Sometimes the insulating indium oxide films were prepared by DC magnetron sputtering for creating tunnel junction barrier layers. Here, we present the development of pure thin In_2O_3 films by magnetron sputtering based on a mathematical model of the process built using RSM.

The experiments were carried out on a laboratory magnetron sputtering vacuum station, equipped with a diffusion pump. A principle schematic diagram of the magnetron sputtering source is shown in Fig. 3.20(a). The residual pressure in vacuum pressure was less than $4 \cdot 10^{-5}$ Torr and the sputtering was done in a pure argon atmosphere (99.996%). The substrate temperature was varied during the deposition process and the samples underwent a post-deposition heat treatment while still in vacuum. The indium oxide films were deposited on a borosilicate glass plate 0.13–0.17 mm thick and on an optical glass slide of 1 mm thickness. The sputtering target was in the shape of a round disk, 50 mm in diameter and 3 mm thick, made of pure indium oxide (99.999%). The substrate-to-target distance was taken as 5 cm.

Optical and electrical properties of the resultant In_2O_3 films were evaluated after preparation. The optical transmittance and absorption for the range of 250/880 nm was measured using a Uvicon 941 Plus spectrophotometer. Transmittance in the IR spectrum was measured in the range of 400/2,500 nm. A computerized metallurgical microscope with magnification ×80 to ×1,600 was used for the surface topography evaluation. The microhardness of the resulting films was measured using a tester PMT-3. The indentation period was 15 s; 5–10 indentations were taken for each specimen. Before deposition, the substrates were cleaned with alcohol in the ultrasound heater for 5 min.

The most significant properties of In_2O_3 films are their transparency and resistivity. These properties were defined as the response functions. The active technological parameters for the sputtering model and their variable ranges were taken as follows: $P_{Ar} = 1.5/2$ Torr is the argon pressure during deposition process, $Vt = 500/750$ V is the DC voltage applied to the target, $T_s = 150/50$°C is the substrate temperature during the deposition process and $t = 20/40$ min is the deposition duration. The model was written in the form: $F = f(P,V,T,t)$. A complete factor experiment of the order of 2^4 yields the following regression equation with 15 coefficients:

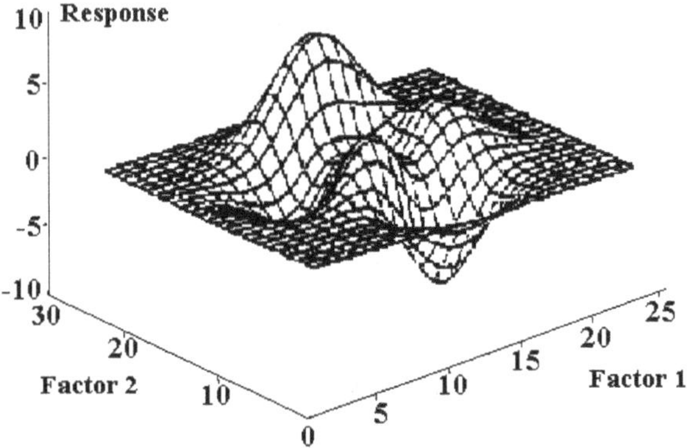

Fig. 4.2 Example of bi-parameter experiment space.

Table 4.3 Normalized active technological parameters.

N	P^*	V^*	T^*	t^*	D	R_s
1	+1	+1	+1	+1		
2	+1	+1	+1	−1		
16	−1	−1	−1	−1		

$$F = a_1 + a_2P_{Ar} + a_3V_t + a_4T_s + a_5t + a_6P_{Ar}V_t + a_7P_{Ar}T_s + a_8P_{Ar}t + a_9V_tT_s + a_{10}V_tt + a_{11}T_st + a_{12}P_{Ar}^{\ 2} + a_{13}V_t^2 + a_{14}T_s^2 + a_{15}t^2 \quad (4.10)$$

Figure 4.2 illustrates an example of a bi-parameter space. The coordinate axes X and Y represent two active technological parameters. Axis Z, in its turn, represents the response function, which can also be named "a transfer function". The response function draws a surface in the multi-parameter space, called the response surface. The response function can be studied with the help of RSM and then optimized. A standard orthogonal plan for this experiment can be presented in a tabular form (see Table 4.3), which is also convenient for numerical calculations.

This table contains the selected normalized active technological factors: P^*, V^*, T^*, t^* and the defined response function D is the transmittance (%) at $\lambda = 550$ nm. The surface resistance R_s (Ω/sq.)

was chosen as a limitation parameter. The normalization was done using Eq. (4.5).

Evidently, in order to create a complete 15-constant "parameter space" from a four-factor system, 15 separate experiments should be carried out. These constants will be then calculated from the 15 obtained equations. In addition, it is necessary to perform a few more experiments in the system center in order to obtain an estimation of the relative experiment error. A virtual example of a multi-parameter space experiment and a random partial section of it is shown in Fig. 4.3. Each one of the experiments is actually a partial section of the chosen multi-parameter space.

These random partial sections are described by the same equations and belong to the analytical response function. Consequently, the coefficients of the approximation model must be equal for all partial sections. Thus, the missing coefficients can be calculated by varying one parameter using Eq. (4.6). The rest of the missing coefficients describing the required model are then obtained by solving the coefficient matrix. One should always bear in mind that only the

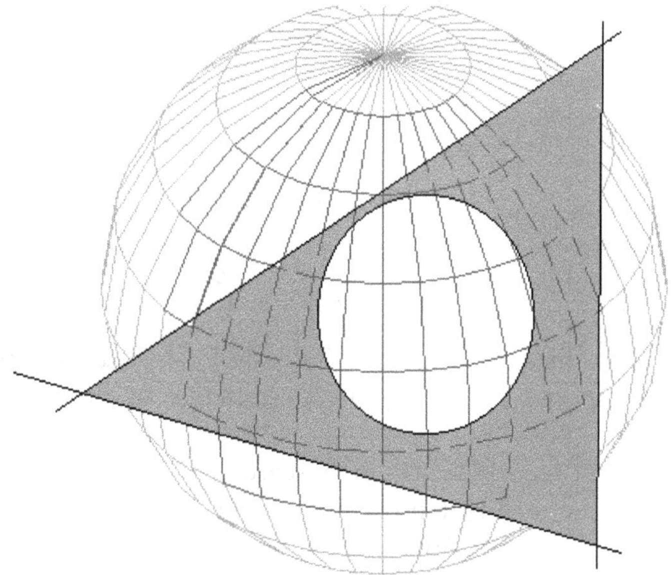

Fig. 4.3 Illustration of random partial sections of the multi-parameter space.

Table 4.4 Experimental measured results.

V (V)	V*	P (Torr)	P*	T (°C)	T*	t (min)	t*	$R_{\#}$	D (%) (550)
400	–1.8	2	1	20	–3.6	30	0	463714.2	90.7
700	0.6	2	1	150	–1	30	0	5142.007	86.02
750	1	1.83	0.32	250	1	30	0	5176.905	88.97
750	1	1.67	–0.32	250	1	20	–1	15391.73	90.9
600	–0.2	1.96	0.84	150	–1	40	1	10656.96	91.9
500	–1	1.5	–1	150	–1	40	1	4706.079	93.47

first few resultant coefficients were experimentally obtained, while the rest of the coefficients were derived mathematically from the following matrix (i is an index):

$$(1\ P_i V_i T_i t_i P_i V_i P_i T_i P_i t_i V_i T_i V_i t_i T_i t_i P_i^2 V_i^2 T_i^2 t_i^2) = (a_i) \cdot (F_i) \quad (4.11)$$

Experiments with random sections in the designed multi-factors space were carried out. Data from these experiments are shown in Table 4.4. Based on these results, the required coefficients were calculated using the model (4.6) for best transmittance and sheet resistance. The designed partial models for the D parameter response (transmittance) are as follows:

$$\begin{cases} D = 89.77 - 3.72P + 0.04P^2 \\ D = 90.83 - 6.04V - 3.37V^2 \\ D = 86.78 + 1.47T + 0.71T^2 \\ D = 86.02 + 0.5t + 5.3t^2 \end{cases} \quad (4.12)$$

According to these expressions, Table 4.5 is filled to give the calculated response parameter D (transmittance at the wavelength $\lambda = 550$ nm).

The model's coefficients, in a regression equation form, are obtained by the standard procedure of the least squares method, taken from Table 4.5. In this calculation, second-order interference effects are considered as well. The required expression is obtained then in the following form:

Table 4.5 Results of the complete factor experiment.

P^*	V^*	T^*	t^*	D
+1	+1	+1	+1	87.09
+1	+1	+1	−1	86.27
+1	+1	−1	+1	90.02
+1	+1	−1	−1	89.29
+1	−1	+1	+1	86.75
+1	−1	+1	−1	86.02
+1	−1	−1	+1	89.77
+1	−1	−1	−1	89.04
−1	+1	+1	+1	88.86
−1	+1	+1	−1	88.14
−1	+1	−1	+1	91.88
−1	+1	−1	−1	91.15
−1	−1	+1	+1	88.61
−1	−1	+1	−1	87.88
−1	−1	−1	+1	91.63
−1	−1	−1	−1	90.9

$$\begin{aligned} F = {} & 88.96 - 0.93\boldsymbol{P} + 0.13\boldsymbol{V} - 1.5\boldsymbol{T} + 0.37\boldsymbol{t} + 0.005\boldsymbol{PV} \\ & + 0.005\boldsymbol{PT} + 0.006\boldsymbol{Pt} + 0.006\boldsymbol{VT} + 0.005\boldsymbol{Vt} + 0.005\boldsymbol{Tt} \\ & + 0.38\boldsymbol{P}^2 + 0.006\boldsymbol{V}^2 + 0.006\boldsymbol{T}^2 + 0.005\boldsymbol{t}^2 \end{aligned} \tag{4.13}$$

Equation (4.13) represents the model of the deposition process in the taken intervals of the input factors. Our following task is to find the optimal values of the input factors providing the maximum value of the response function satisfying the given limitations.

4.3 Optimization of the thin film deposition process

Figure 4.4 illustrates the graphical interpretation of the response function $f(x,y)$.

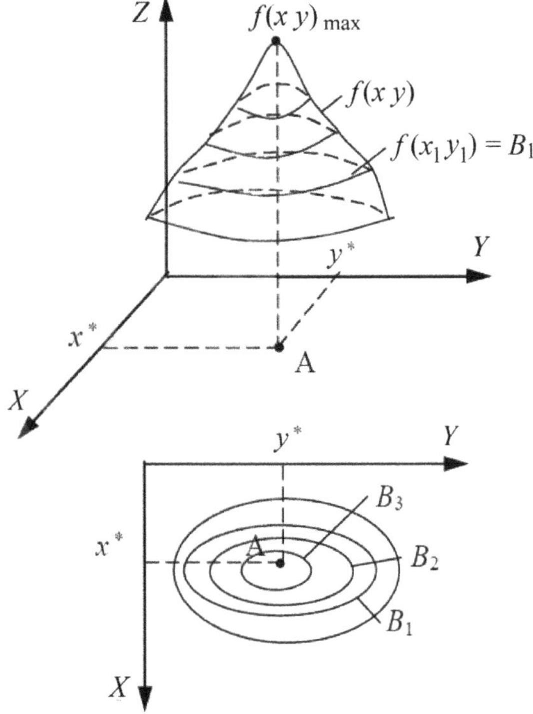

Fig. 4.4 Graphical representation of the optimization of function $f(x, y)$.

Here, the point A matches the optimum values of the factors x^* and y^* providing the maximum of the response function. The closed lines in Fig. 4.4 characterize lines of constant level and are described by the equation $f(x,y) = B = \text{const}$. Search methods for optimal values belong to the class of iterative procedures, while the whole process is divided into steps, at each step a number of experiments are carried out and it is determined how to change factors influencing the process to get an improvement in the result. Wherein at each successive step, the information obtained is used to select the next step. There are many methods for iterative optimization. We will consider some of them here.

The simplex planning method allows you to find the optimum area without preliminary study of the influence of factors. This method does

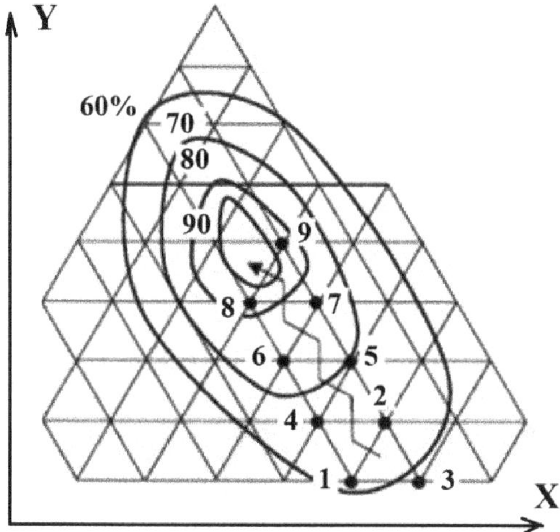

Fig. 4.5 Moving to the optimal area using the simplex method.

not require the calculation of the gradient of the response function; therefore, it belongs to the gradient-less methods for finding the optimum. A simplex is the simplest convex polyhedron formed by k+1 vertex in k-dimensional space, which are connected by straight lines. In this case, the coordinates of the vertices of the simplex are the values of the factors in individual experiments: in a two-factor space, it is any triangle, in a three-factor space — tetrahedron. Figure 4.5 represents a geometric image of a simplex for a two-dimensional case.

After designing the initial simplex and providing experiments with values of the factors corresponding to the coordinates of its vertices, the results are analyzed and the vertex of the simplex is selected, from which the lowest (worst) value of the response function is obtained. To move to the optimum, it is necessary to set an experiment at a new point, which is a mirror image of the point with the worst result relative to the opposite face of the simplex. For example, according to the results of tests 1, 2, 3, experience 3 turned out to be the worst. The next test is placed at point 4, which forms a new regular simplex with points 1 and 2. Further, the results of experiments 1, 2, 4 are compared. The worst result is obtained at point 1, so it is replaced in the simplex by a specular reflection (point 5), and so on, until an almost stationary region is reached. Thus, a zigzag path is turned out. The total number of tests required to reach the optimum region may be small due to the fact that $k+1$ test has to be carried out only at the beginning, and then each step is accompanied by only one additional test, the conditions of which are selected on the basis of the previous results.

Another way to reach the optimal combination of input factors is moving in the direction of the function's gradient slope. This method is called the "steep rise method". According to this method, the moving should be done at the direction normal to the lines of equal level. The function optimum (extreme) is achieved under the following conditions:

$$\nabla f(X_1 X_2 \ldots X_m) = 0 \tag{4.14}$$

and

$$\nabla^2 f(X_1 X_2 \ldots X_m) < 0 \text{ (for the maximum)} \tag{4.15}$$

Figure 4.6 illustrates the steep rise method.

To move along the gradient of the response function, it is necessary to change the regression coefficients in proportion to their

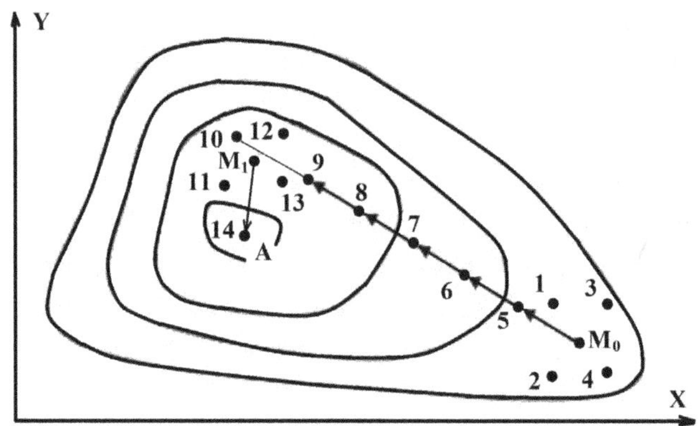

Fig. 4.6 Steep rise method illustration.

regression coefficients and in the direction corresponding to the sign of the coefficient. The movement continues until reaching the local extremum. Now, the process should be repeated up to achieving the global maximum.

If we return now to the model of the deposition process (see Eqs. (4.11) and (4.13)), we can take the gradient of the response function according to Eq. (4.14):

$$\partial F\left(P^*,V^*,T^*,t^*\right)=\nabla F(P,V,T,t)\cdot\begin{pmatrix}\partial P\\ \partial V\\ \partial T\\ \partial t\end{pmatrix}=0 \qquad (4.16)$$

where $F(P^*,V^*,T^*,t^*)$ is the required extremum point. Analysis of this relation yields 16 possible combinations (2^4) between the four variables. Similarly, analysis of the second derivative yields another type of extremum (maximum or minimum). Calculations of the optimal deposition parameters were done using the "steep rise" method (Eq. (4.16)). Optimization was done for two parameters: transmittance at 550 nm and sheet resistance. Resultant optimized technological values for these two parameters are shown in Table 4.6.

A typical transmittance characteristic of obtained In_2O_3 thin films is shown in Fig. 4.7.

Prepared indium oxide thin films have a good adhesion to the glass and a very homogeneous structure with grain dimensions smaller than the optical microscope resolution. The calculated optical bandgap of the films was approximately 3.8 eV, which is in agreement with reference sources.

Table 4.6 Optimized technological parameters (input factors and the response function).

V_t, (V)	P_{Ar} (Torr)	T_s (°C)	t (min)	d (Å)	$R_\#$ (Ω/sq)	D (%) (550 nm)
1000	2.03	172	15	720	2438.22	88.23
750	1.76	179	34	2525	1704.03	90.97

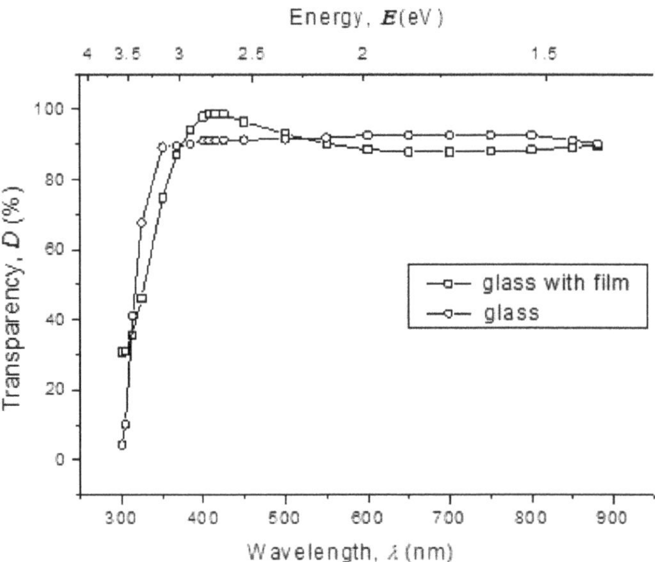

Fig. 4.7 Typical transmittance characteristic of In_2O_3 thin film on borosilicate glass.

The model error for Eq. (4.12) was evaluated at the point (V, P, T, t) = (−1, −1, −1, 1). The relative error in the selected domain of input active factors was equal:

$$\delta = \frac{|\Delta|}{F_{(-1,-1,-1,1)}} 100\% = \frac{93.47 - 91.63}{91.63} 100\% = 2\% \qquad (4.17)$$

This small error is therefore the proof that the proposed model is adequate and may be used for setting magnetron deposition parameters. The designed model's coefficients (regression coefficients) may be presented in Taylor series of coefficients around the points of interest. This approximation method is therefore most convenient since accuracy of the solution system can be enhanced by increasing Taylor equation's degree.

The main disadvantage of this model lies in the quadratic approximation for multilayer processes due to the inaccuracy of the linear approximation there. It should be noted that this approximation does not precisely represent the given mathematical function since the real function may be too complicated to deal with.

4.4 Measurement and processing errors

Measurement is the process of experimentally finding the value of the physical quantities using specific technical means. Unfortunately, no measurement can be made with absolute precision. As a result, we always get the value of the quantity with some error. Therefore, the task of measurements includes not only finding the values of the quantities, but also estimation of the error allowed for this. *Measurement error* is the deviation of the measured value from the true value of the measurement. In this case, a distribution is made between *absolute* and *relative* errors. The absolute measurement error is the difference between the measured x and the true x_t values of the measured quantity, expressed in the unit of the measured quantity:

$$\Delta x = x - x_t \tag{4.18}$$

The relative measurement error is the ratio of the absolute error to the true value of the measured value, expressed in percentages:

$$\delta_x = \frac{\Delta x}{x_t}\% \tag{4.19}$$

Measurement error can be due to many different factors, one of the significant ones being an instrumental error that is defined by the resolution of the measuring devices. Moreover, the true value of the studied physical parameter is usually not known. Thus, we can define it approximately, using Eq. (4.18), taking into account the instrumental error (usually the device's resolution) as the absolute error. A reliable and scientifically grounded method for determining the coefficients of experimental dependencies is the least squares method. Its essence is in the selection of such values of the coefficients at which the sum of the squares of the deviations of the values f_i $(i = 1, 2,\ldots,n)$ measured in the experiments from the calculated ones would be minimal. The least squares method allows not only finding the coefficients of the functional dependence, but also estimating the errors of the found coefficients.

Let us consider the least squares method with the following assumptions:

- there is a dependence $y = a + bx$ between two variables x and y;
- deviations of x are small.

To calculate the coefficients a and b, it is necessary to provide the following sequence of operations:

1. Experimentally get n pairs values of the argument x and function y, (x_i, y_i).
2. Calculate average values of all experimental points:

$$x_{av} = \frac{1}{n}\Sigma x_i; \quad y_{av} = \frac{1}{n}\Sigma y_i \tag{4.20}$$

3. Find the functional coefficients according to the following equations:

$$b = \frac{\Sigma(x_i - x_{av})(y_i - y_{av})}{\Sigma(x_i - x_{av})^2}; \quad a = y_{av} - bx_{av} \tag{4.21}$$

4. Calculate additional parameters D and d_i:

$$d_i = y_i - (a + bx_i); \quad D = \Sigma(x_i - x_{av})^2 \tag{4.22}$$

5. Calculate the average square deviations of the coefficients a and b (see also Eq. (4.8)):

$$S_b^2 = \frac{1}{D}\frac{\Sigma d_i^2}{n-2}; \quad S_a^2 = \left(\frac{1}{n} + \frac{x_{av}^2}{D}\right)\frac{\Sigma d_i^2}{n-2} \tag{4.23}$$

6. The casual error of the calculated function y for an arbitrary argument x is calculated by the following equations at the confidence level $\alpha = 0.95$ (A confidence level of 0.95 means that there is a probability of at least 95% that the result is reliable):

$$\Delta_y = \sqrt{(2S_a)^2 + (2S_b x)^2} \tag{4.24}$$

Evidently, due to the relatively large amount of calculations, it is advisable to carry out the least squares method with the help of suitable computer programs. Also, the type of the expected relations between variables should be taken into account. In the case of a regression equation of a polynomial type, calculation of the

Table 4.7 Errors of functions of one variable.

$z = z(x)$	Δ_z	δ_z
cx, c = const	$c\Delta_x$	$\frac{\Delta_x}{x}$
x^n, $n \leq 0$	$nx^{(n-1)}\Delta_x$	$n\frac{\Delta_x}{x}$
$\frac{c}{1+x}$	$\frac{\Delta}{(1+x)^2}$	$\frac{\Delta_x}{c(1+x)}$
$\frac{c}{1-x}$	$\frac{\Delta_x}{(1-x)^2}$	$\frac{\Delta_x}{c(1-x)}$
$\sqrt[n]{x}$	$\frac{1}{n}\frac{\sqrt[n]{x}}{x}\Delta_x$	$\frac{1}{n}\frac{\Delta_x}{x}$
$e^{\frac{x}{c}}$, c = const	$e^{\frac{x}{c}}\frac{\Delta_x}{c}$	$\frac{\Delta_x}{c}$
$A^{\frac{x}{c}}$, c = const, A = const	$A^{\frac{x}{c}}\ln A\frac{\Delta_x}{c}$	$\ln A\frac{\Delta_x}{c}$
$\ln x$	$\frac{\Delta_x}{x}$	$\frac{\Delta_x}{x\ln x}$
$\sin\left(\frac{x}{c}\right), c = \text{const}$	$\cos\left(\frac{x}{c}\right)\frac{\Delta_x}{c}$	$\text{ctg}\left(\frac{x}{c}\right)\frac{\Delta_x}{c}$
$\cos\left(\frac{x}{c}\right), c = \text{const}$	$\sin\left(\frac{x}{c}\right)\frac{\Delta_x}{c}$	$\text{tg}\left(\frac{x}{c}\right)\frac{\Delta_x}{c}$
$\text{tg}\left(\frac{x}{c}\right), c = \text{const}$	$\frac{1}{\cos^2(x/c)}\frac{\Delta_x}{c}$	$\frac{2}{\sin(2x/c)}\frac{\Delta_x}{c}$
$\text{ctg}\left(\frac{x}{c}\right), c = \text{const}$	$\frac{1}{\sin^2(x/c)}\frac{\Delta_x}{c}$	$\frac{1}{\sin^2(2x/c)}\frac{\Delta_x}{c}$

coefficients will be other than Eq. (4.21). Table 4.7 represents the calculated formulas for errors of the function of one argument.

It should be noted that during the development of the technological deposition of thin films with given properties, we measure two types of quantities: independent acting input factors (technological parameters) and properties of deposited thin films. All these measurements are implemented with different measuring devices, each of them has a different resolution and precision. This requires specific attention as much at the stage of the model's design as at the stage of the analysis of the built model.

Chapter 5

Optical properties and thickness of thin films

5.1 Optical properties of thin films

5.1.1 *Introduction*

Light is a type of energy that propagates into space at a very high speed. More precisely, light is understood as an electromagnetic wave traveling into space — it is radiant energy. Figure 5.1 represents the spectrum of electromagnetic waves when the visible spectrum takes a suitable part.

Interaction between an electromagnetic wave and a matter brings information on the composition, the structure, the optical and electrical properties of the matter. In dependence with the energy of the electromagnetic waves, different properties may be evaluated. Here, we consider interaction between electromagnetic waves in several selected ranges with thin films of different materials.

When light passes through a thin film, light energy is dissipated according to the law of energy conservation:

$$I_0 = I_R + I_A + I_T \tag{5.1}$$

where I_0 is an intensity of the incident light, I_R is an intensity of the reflected light, I_A is intensity of light absorbed in the thin film and

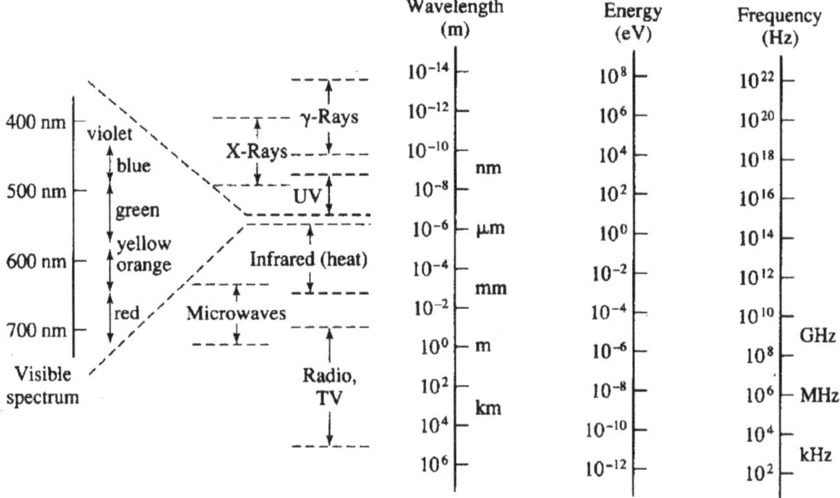

Fig. 5.1 Electromagnetic waves spectrum.

I_T is an intensity of the transmitted light. If we denote following parameters $R = I_R/I_0$ as the reflectance, $A = I_A/I_0$ as an absorbance and $T = I_T/I_0$ as a transmittance, Eq. (5.1) transforms to the simplified equation which may be estimated in percents:

$$R + A + T = 100\% \tag{5.2}$$

Figure 5.2 illustrates propagation of light through the material film with a thickness d.

As shown in Fig. 5.2, an incident light I_0 while interacting with a thin film undergoes the following processes: A part of the energy of light I_R reflects from the film surface and returns to space, a second part I_A enters into the film and is absorbed there, and a third part I_T passes through the film. It should be noted that the light intensity decreases inside the materials exponentially according to the extent α, known as an absorption coefficient, and the thickness of the material. To describe all effects of the interaction between light and matter, the known Maxwell equations in the differential form should be applied (see Eq. (3.50)):

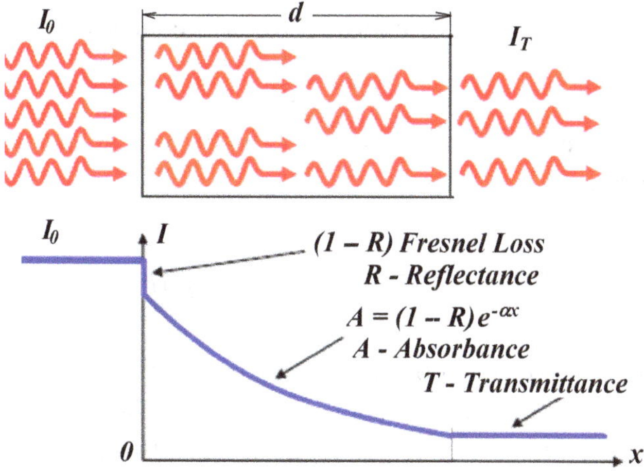

Fig. 5.2 Propagation of light though a thin film.

$$\begin{cases} \nabla \cdot \boldsymbol{E} = \dfrac{\rho}{\varepsilon_0} (\text{Gauss's law}) \\ \nabla \cdot \boldsymbol{B} = 0 (\text{Gauss's law for magnetism}) \\ \nabla \times \boldsymbol{E} = -\dfrac{\partial \boldsymbol{B}}{\partial t} (\text{Faraday's law for induction}) \\ \nabla \times \boldsymbol{B} = \varepsilon_0 \mu_0 \dfrac{\partial \boldsymbol{E}}{\partial t} + \mu_0 \boldsymbol{j} (\text{Ampere's circuital law}) \end{cases} \tag{5.3}$$

however now, equations of matter should be taken into account:

$$\begin{cases} \boldsymbol{D} = \varepsilon \boldsymbol{E} + \boldsymbol{P} \\ \boldsymbol{B} = \mu \boldsymbol{H} + \boldsymbol{M} \\ \boldsymbol{j} = \sigma \boldsymbol{E} \text{ Ohm's law} \\ c = \dfrac{1}{\sqrt{\varepsilon_0 \mu_0}} \text{in vacuum} \end{cases} \tag{5.4}$$

where *D* is the electrical displacement, *P* is the polarization, B is the magnetic induction, *M* is the magnetization, ε_r and μ_r are relative permittivity and permeability of the film's material consequently, *j* is the current density, σ is the conductivity of the material and $c_m = \sqrt{\varepsilon\mu}$ is the light velocity in the material. Here, $\varepsilon = \varepsilon_0\varepsilon_r$ and $\mu = \mu_0\mu_r$. If we compare velocities of light in vacuum and in the matter, one can see that the light velocity in the matter depends on the matter parameter called the refraction coefficient:

$$n \equiv \frac{c}{c_m} = \frac{\lambda \cdot t}{\lambda_m \cdot t} = \frac{\sqrt{\varepsilon\mu}}{\sqrt{\varepsilon_0\mu_0}} = \sqrt{\varepsilon_r\mu_r} \tag{5.5}$$

where λ is the wavelength of light in vacuum and λ_m is the wavelength of light in the material. As shown in Eq. (5.5), the refraction coefficient value depends on the wavelength of light; this property is called the dispersion of light. Due to the law of energy conservation, when light enters into the matter from vacuum, its velocity and wavelength decrease to keep the energy of photons that is directly proportional to the frequency to be constant.

Let us combine equations of Faraday and Ampere (see Eq. (5.3)). First, we take a rotor from the Faraday equation:

$$\nabla \times (\nabla \times \boldsymbol{E}) = \nabla \times \left(-\mu \frac{\partial \boldsymbol{H}}{\partial t} \right) \tag{5.6}$$

The left-hand side of this equation, considering that $\nabla(\nabla E) = 0$, will take the form

$$\nabla \times (\nabla \times E) = \nabla(\nabla E) - \nabla^2 E = -\nabla^2 E \tag{5.7}$$

The right-hand side of Eq. (5.6), considering Ohm's law, transforms to

$$\nabla \times \left(-\mu \frac{\partial H}{\partial t} \right) = -\mu \frac{\partial}{\partial t}(\nabla \times H) = -\mu \frac{\partial}{\partial t}\left(j + \varepsilon \frac{\partial E}{\partial t} \right) \tag{5.8}$$

Combining Eqs. (5.7) and (5.8), we get the equation describing the light propagation in a magnetic and absorbing environment:

$$\nabla^2 E = \mu\varepsilon \frac{\partial^2 E}{\partial t^2} + \mu\sigma \frac{\partial E}{\partial t} \tag{5.9}$$

Solution of this equation represents waves with decreasing amplitude:

$$E = E_0 e^{-\frac{\omega\kappa}{c}x} e^{\left[i\omega\left(t-\frac{nx}{c}\right)\right]} \tag{5.10}$$

where $\omega = 2\pi\nu = 2\pi/\lambda_m$ is the angular frequency of light, ν is the frequency of light, k is the extinction coefficient providing decay of the electromagnetic field amplitude. If we insert the solution of (5.10) into the wave-equation (5.9), we get the following significant equation:

$$n^2 = \varepsilon - i\frac{\sigma}{\omega} = \varepsilon - i\frac{\sigma}{2\pi\nu} \tag{5.11}$$

So, analysis of the Eq. (5.11) shows: Electrical and optical properties of thin films' materials are closely related:

$$\varepsilon = n^2 - k^2; \ \sigma = 4\pi nk\nu \tag{5.12}$$

Also, the main parameters of materials such as the refraction coefficient and permittivity are complex and have real and imaginary parts, all being main characteristics of materials under light illumination dependent on the wavelength of light and they may be called spectral.

The intensity of light represents the strength of an electric field in a wave of light $I = E^2$. Each material more or less absorbs the incident light. The typical light penetration depth w in the material is defined by the intensity decreasing to e^{-1} times. A value that equals to $\alpha = w^{-1}$ is called the absorption coefficient:

$$\alpha = \frac{1}{w} = \frac{4\pi\kappa}{\lambda} = \frac{4\pi\sigma}{nc} = \frac{2\omega\kappa}{c}\left[\mathrm{cm}^{-1}\right] \tag{5.13}$$

Figure 5.3 represents, for example, the absorption coefficients for different semiconductor materials. Absorption coefficient is a very significant parameter of thin films and specifically semiconductor thin films. Further, we will discuss about measurement of this

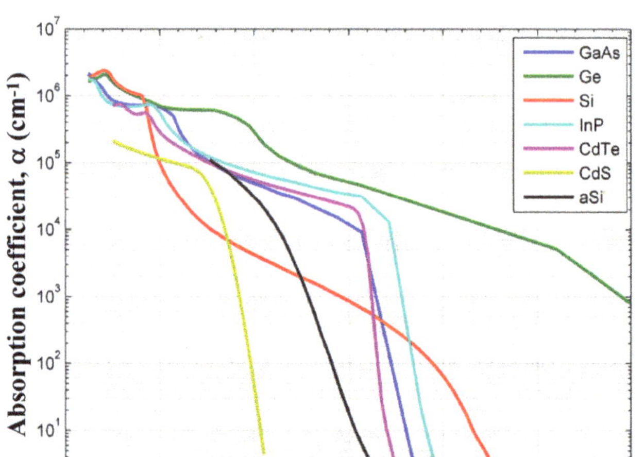

Fig. 5.3 Absorption coefficient for different semiconductor materials.

parameter in real thin films. The physical meaning of the absorption coefficient is a probability that a photon with the given energy will be absorbed.

$$\alpha = \frac{1}{I(h\nu)} \frac{dI(h\nu)}{dx} \tag{5.14}$$

The mechanism of absorption is complex especially in the case of semiconductor thin films. There are several different mechanisms of absorption defined by energetic structure of the material, its crystal structure and presence of defects:

- Self-own absorption consisting of absorption of photons with energy enough to transfer electrons from the valency band to the conductive band.
- When there is not enough energy, there is excitation of electron with formation of excitons (not stable quasi-atom of hydrogen).
- Impurity atoms form additional energetic levels trapping the excited electrons.

- Sometimes, excitation leads to transfer of the electrons inside the same energy band.
- Phononic absorption, that is trapping photons by the lattice vibrations.

Detailed consideration of all these types of absorption is beyond the scope of this book. It should be noted that absorption can happen only if the photon energy is more or equal to the bandgap of the semiconductor. So, there is a red border for each semiconductor material described by the following formula:

$$\lambda_r = \frac{1.24}{E_g} \tag{5.15}$$

where λ_r is the red border in μm and E_g is the bandgap energy in eV. The semiconductor film will be transparent for photons with energy less than the bandgap. This claim is illustrated in Fig. 5.3: A red border exists for each semiconductor. Also, a bandgap of the semiconductor may be found from spectral transmittance characteristics of semiconductor thin films.

5.1.2 *Introduction to spectrometry*

Optical spectroscopy is based on the interaction of light with matter. The main properties of materials may be determined using analysis of light absorbed or emitted by a sample. For this goal, we use light of the near ultra-violet, visible and infrared ranges. For example, the light absorption can be used in analytical chemistry to characterize and determine the composition and amount of substances. Figure 5.4 represents two colored objects.

Both vegetables shown in Fig. 5.4 are illuminated by visible or white light, represented by a rainbow: The different colors represent the different components of visible light. When light rays illuminate an object, they may be absorbed by the object — in particular, one or more light components (i.e., its colors) are specifically absorbed.

Fig. 5.4 Two colored objects illuminated by white light.

The colors that are not absorbed by the object are reflected. Practically, spectroscopy is the measurement and recording of transmittance, reflectance or absorbance characteristics of a material as a function of wavelength.

Compared with other analytical methods, optical spectroscopy has the following advantages:

- The optical spectroscopy method is non-destructive and not aggressive.
- These measurements may be provided from any distance without physical contact with the measured sample.
- In general, for the optical spectroscopic measurements, the transmittance of the measured sample does not matter and nor does its physical state: solid, liquid or gaseous.

There are many more specific spectroscopic techniques which are widely used in the world. Here, we will consider the optical spectroscopy in the UV-VIS range and the evaluation of materials in the far infrared range.

5.1.3 *Spectrometry in the UV-VIS range*

UV-VIS spectroscopy is a technique based on the absorption of light by an unknown substance or by an unknown sample. Here, the sample is illuminated by electromagnetic rays at different visible wavelengths (the different colors) and adjacent ranges, UV and near IR of

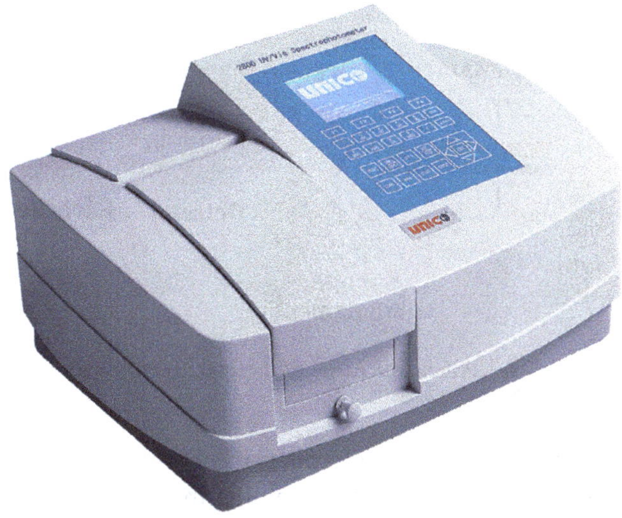

Fig. 5.5 The spectrophotometer for UV-VIS range.

the spectrum, i.e., 180–1,200 nm range. Depending on the material, a part of the light is partially absorbed. The remaining light, i.e., transmitted light, is recorded as a function of wavelength by a suitable detector, which provides the UV-VIS spectrum of the sample. Figure 5.5 represents the spectrophotometer which measures the transmittance and absorbance characteristics in the range of 180–1,100 nm.

The apparatus shown in Fig. 5.5 consists of five parts: (1) Halogen or deuterium lamps to supply the light; (2) A Mono-chromator to isolate the wavelength of interest and eliminate the unwanted second-order radiation; (3) A sample compartment to accommodate the sample solution, if the sample is in the liquid state, or an arrangement for the solid sample (a thin film on the suitable substrate); (4) A detector to receive the transmitted light and convert it to an electrical signal; and (5) A digital display to indicate absorbance or transmittance. Figure 5.6 that follows illustrates the relationship between these parts. In the spectrophotometer, light from the lamp is focused on the entrance slit of the monochromator where the collimating mirror directs the beam onto the grating. The grating disperses the light

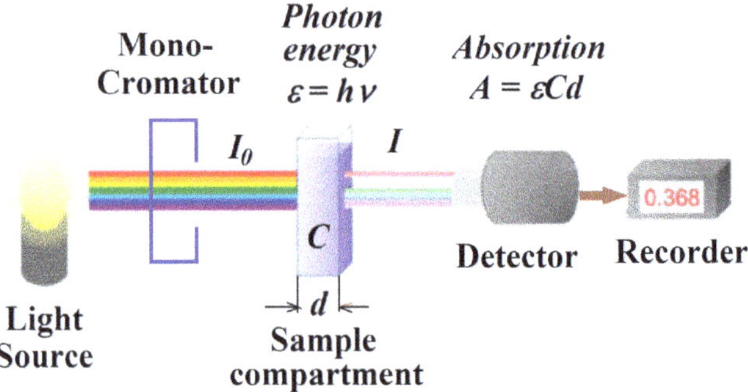

Fig. 5.6 Principle scheme of the spectrophotometer.

beam to produce the spectrum, a portion of which is focused on the exit slit of the monochromator by a collimating mirror. From here the beam is passed to a sample compartment through one of the filters, which helps to eliminate unwanted second-order radiation from the diffraction grating. Upon leaving the sample compartment, the beam is passed to the silicon photodiode detector and causes the detector to produce an electrical signal that is displayed on the digital display. Also, this signal can be transferred to the computer to record and remember it. An absorption of light passing through a liquid cell is shown in Fig. 5.6. Here, a liquid solution containing a studied material with concentration C is placed in the glass or quartz cuvette with thickness d. Thus, an absorption will be described by the following formula:

$$A = \varepsilon C d \tag{5.16}$$

where $\varepsilon = h\nu$ is the energy of the photon and d is the cuvette dimension.

Spectral characteristics recorded using the spectrophotometer and processed appropriately allow us to obtain the required parameter of the studied substance. For example, let us consider two types of gold thin films: Solid and non-continuous, as shown in Fig. 5.7.

We discussed earlier (see Chapter 3) how a thin film is generated and grown. Firstly, the critical dimension clusters appear on the surface. They move on the substrate surface, coalesce and finally form a

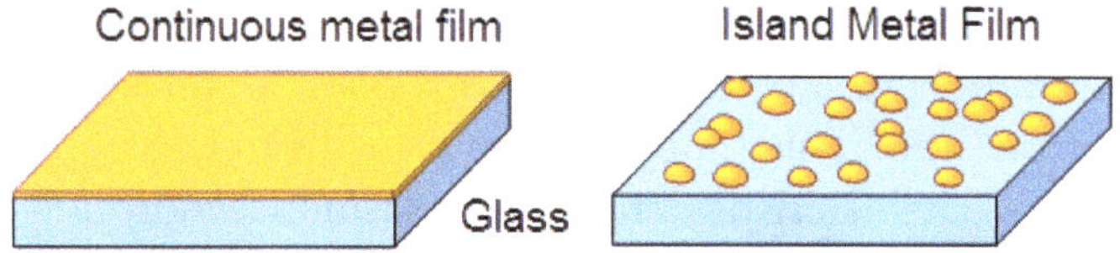

Fig. 5.7 Two different types of thin films: Continuous and island.

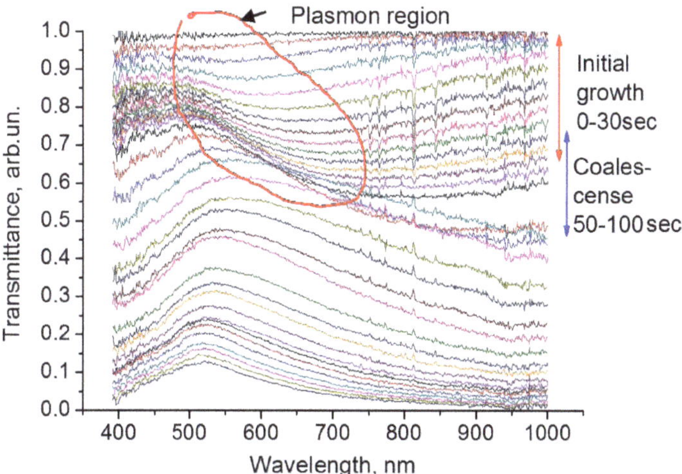

Fig. 5.8 Transmittance characteristics of the sputtered gold film recorded in-situ during growth.

continuous thin film. Thus, Fig. 5.7 illustrates two stages from the film growth. However, the island (non-continuous) thin films also find their applications sometimes. The point is that on small metal islands of some metals, such as gold, silver, copper, under light illumination, a specific effect called the plasmon appears. It consists of polarization of these small islands, moreover a hard electromagnetic field is generated by them. The energy and frequency of this field are defined by dimensions of islands and their distribution on the surface of the substrate. The islet's material comes into resonance with the incident light and absorbs the light of the specific frequency.

Figure 5.8 represents the series of consequent transmittance characteristics measured and recorded through the sputtering deposition process of a thin gold film on the glass substrate.

Figure 5.8 shows the change in the transmittance of the growing thin gold film. Through the first ~30 s, we are looking at the initial growth of the film. This time, the film remains of islet type and the transmittance characteristics show the plasmon absorption valleys. Due to the coalescence process, islands come closer and connect with eachother. After that, we see the continuous solid film without additional absorption peaks. From Fig. 5.8 we can see the characteristic transparency peak for gold at about 500 nm. It can be seen that the plasmon absorption wavelength depends on the film thickness. We can compare the transmittance characteristics related to different film thicknesses.

Another example is in estimation of the optical bandgap of semiconductor thin films using transmittance characteristics. For this, we need to analyze the spectral characteristics in the band-edge region. The calculation method is the well-known Tauc method. It is based on the empirical relation between the optical absorption and the difference between the photon energy and the bandgap:

$$\alpha h\nu = B(h\nu - E_g)^n \tag{5.17}$$

where α is the absorption coefficient, $h\nu$ is the photon energy, B is an independent constant coefficient. The value of the exponent n depends on the kind of electronic transition in the semiconductor material. For semiconductors with direct transition, such as GaAs or ZnO, this exponent $n = ½$ or $3/2$; for semiconductors with indirect transition, such as Si or Ge, $n = 2$ or 3. The value of the exponent in each specific case should be selected for the thin film from the condition when the curve gives the best linear shape in the band-edge region. Figure 5.9 represents transmittance characteristics (a) and Tauc's plots (b) of the complex thin film composed from zinc oxide with allowed impurities of europium and ytterbium prepared by the magnetron sputtering method.

The deposition process was provided by simultaneous sputtering from three different targets and the relation between components was defined by the relation of sputtering power applied to the targets. The transmittance characteristics shown in Fig. 5.9(a) represent properties of as-deposited and annealed thin films. As can be seen from the

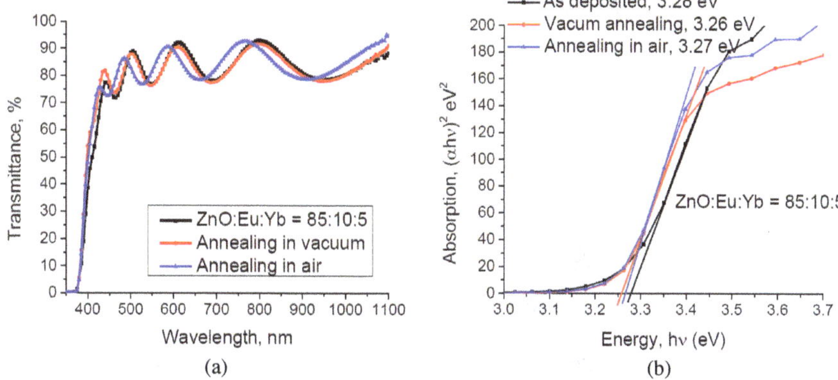

Fig. 5.9 Transmittance characteristics of the complex thin film ZnO:Eu:Yb.

transmittance characteristics, annealing in vacuum slightly reduces the transparency of the film and changes the bandgap. On the other hand, annealing in air leads to an increase in the transparency of the film and slight increase in its thickness. This is due to the additional oxidation of impurities and the precipitation of the corresponding phases. As shown in Fig. 5.9(b), upon annealing of complex films, the optical bandgap changes, which is caused by the formation of additional traps in the bandgap. Since ZnO is a wide-bandgap semiconductor with allowed direct transition between valence and conductive bands during absorption, the exponent was taken as $n = ½$. As is known, the absorption edge of non-metallic materials, crystalline and non-crystalline, is determined by the transition of electrons from the valence band to the conduction band upon absorption of photons with sufficient energy. In this case, the absorption coefficient, that is the number of acts of electron transitions between zones, sharply increases to value of the order of 10^3–10^4 cm^{-1}, depending on the photon energy.

5.1.4 *Spectrometry in the far IR range*

The atoms in various materials at the room temperature oscillate with frequencies in the IR range. The total internal energy of a molecule in a first approximation can be resolved into the sum of rotational,

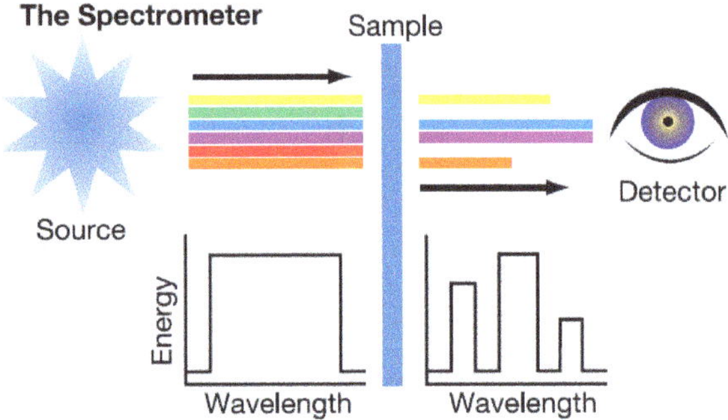

Fig. 5.10 Principle of operation of the IR spectrometer.

vibrational and electronic energy levels. Infrared spectroscopy is the study of interactions between matter and electromagnetic fields in the IR region. In this spectral range, the electro-magnetic waves mainly couple with the molecular vibrations. In other words, a molecule can be excited to a higher vibrational state by absorbing IR radiation. The probability of a particular IR frequency being absorbed depends on the actual interaction between this frequency and the molecule. In general, a frequency will be strongly absorbed if its photon energy coincides with the vibrational energy levels of the molecule. IR radiation is transmitted through a sample as shown in Fig. 5.10, which represents an operation principle of an IR spectrometer. Some part of infrared radiation is absorbed by the sample and a part transfers through it.

The resulting spectrum represents the molecular absorption of the sample. In the field of IR irradiation, the following units are useful: Wavenumbers (cm^{-1}). For example, 1,000 cm^{-1} relates to the wavelength equal to 10 μm and an energy of 1 eV relates to the wave number ~8,100 cm^{-1}. We may also define the application borders: 4,000–14,000 cm^{-1} defines the near-IR region, the mid-IR field is defined by 500–4,000 cm^{-1} and a range 5–500 cm^{-1} represents the far-IR region.

A spectrophotometer working in the field of infrared wavelength is called the Fourier transform infrared (FTIR) spectrophotometer. The interferometer produces a unique type of signal which has all of the infrared frequencies “encoded” into it. The resulting signal is called an

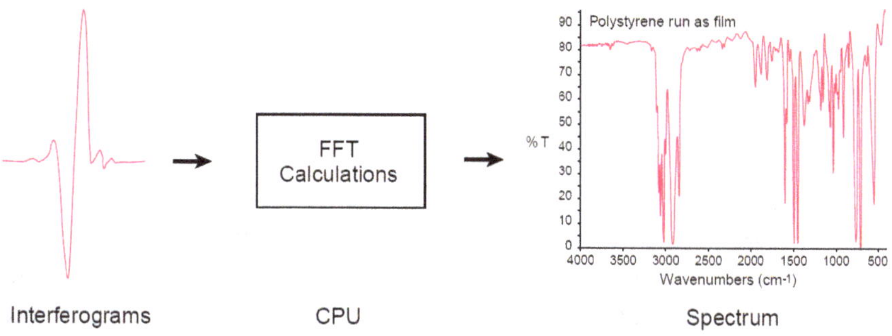

Fig. 5.11 FTIR operation principle.

interferogram. It has the unique property that every data point (a function of the moving mirror position) which makes up the signal has information about every infrared frequency which comes from the source. Figure 5.11 illustrates the principle of operation of FTIR.

Because the analyst requires a frequency spectrum (a plot of the intensity at each individual frequency) in order to make an identification, the measured interferogram signal cannot be interpreted directly. A means of "decoding" the individual frequencies is required. This can be accomplished via a well-known mathematical technique called the Fourier transformation. This transformation is performed by the computer, which then presents the user with the desired spectral information for analysis. Comparison of obtained spectra with the database of various materials allows to determine the composition of the studied material. Also, varying treatment parameters helps to study their influence on the material properties.

One of the important tasks of spectroscopy methods is to understand which frequencies will be absorbed during recording and why. In the usual dispersive-type spectrometer, a grating or a prism is used to disperse light into individual frequencies, and a slit is placed in front of the detector to determine which frequency can reach the detector. However, the FTIR spectrometer operates on a different principle called Fourier transform. The mathematical expression of Fourier transform can be expressed as

$$F(\omega) = \int_{-\infty}^{+\infty} f(x) e^{i\omega x} dx \tag{5.18}$$

And the reverse Fourier transform is

$$f(x) = \frac{1}{2\pi}\int_{-\infty}^{+\infty} F(\omega)e^{i\omega x}d\omega \tag{5.19}$$

where ω is angular frequency and x is the optical path difference in our case. $F(\omega)$ is the spectrum and $f(x)$ is the interferogram. It is clear that if the interferogram $f(x)$ is determined experimentally, the spectrum $F(\omega)$ can be obtained by using Fourier transform. Through the Fourier transform, the interferogram is transformed into IR absorption spectrum that is commonly recognizable with absorption intensity or % transmittance plotted against the wavelength or wavenumber. The ratio of radiant power transmitted by the sample (I) relative to the radiant power of incident light on the sample (I_0) results in quantity of Transmittance, (T). Absorbance (A) is the logarithm to the base 10 of the reciprocal of the transmittance (T). For example, Fig. 5.12 represents the infrared reflectance spectrum of thin vanadium oxide film on the polyimide substrate.

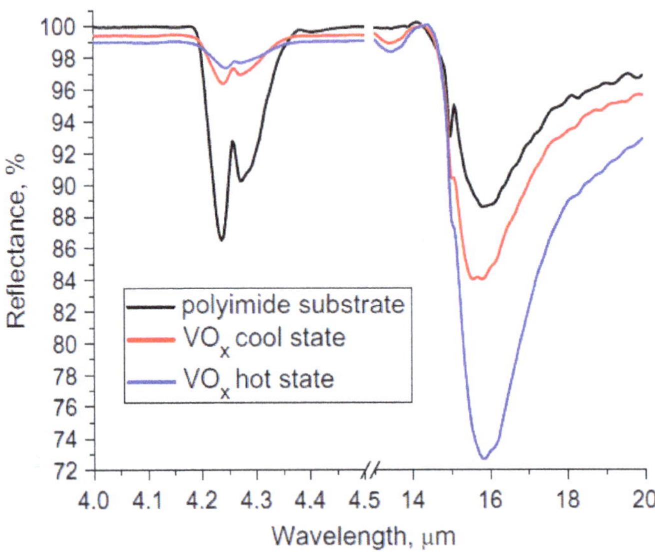

Fig. 5.12 IR reflectance spectrum of VO_x thin films on polyimide substrate.

Vanadium oxide VO_2 is a material that transforms from semiconductor to a metal state at temperature of 67°C. The phase transformation is accompanied by dramatic changes of its electrical and optical properties. Figure 5.12 shows the reflectance spectra of the pure polyimide substrate and the polyimide coated by the thin film consisting of the mixture of different vanadium oxides (VO_x). These films were deposited by the vacuum thermal evaporation method, which cannot provide the stoichiometric composition of the coating; vanadium is the transition metal which has various valencies and it can be used to form various oxides. In this case, infrared spectroscopy helps to explain the changes occurring in the thin film under thermal treatment.

5.2 Thin films' thickness and its characterization

5.2.1 *Introduction*

Thickness of a coating is an important parameter which defines it belongs to the category of thin films. We know already that there is a size-effect influencing the film properties. As shown in Fig. 1.6, a thin film repeats the surface structure of the substrate and the thickness is commensurate with the surface defects. Also, there are noncontinuous thin films consisting of islands, usually shaped by the discs separated from each other. From this, one significant question arises: What is the thickness of a thin film and what do we measure when talking about thickness? This may be an average thickness defined by the mass, density and area occupied by the coating or a maximal height of the islands or a roughness. However, the density of the deposited layer may be different from the density of the solid material. So, the thickness of thin film is closely related with the measurement method and measuring devices, which have different resolutions and precision.

There are many thickness measurement techniques. For example, a profilometry is the straightforward method to measure the coating

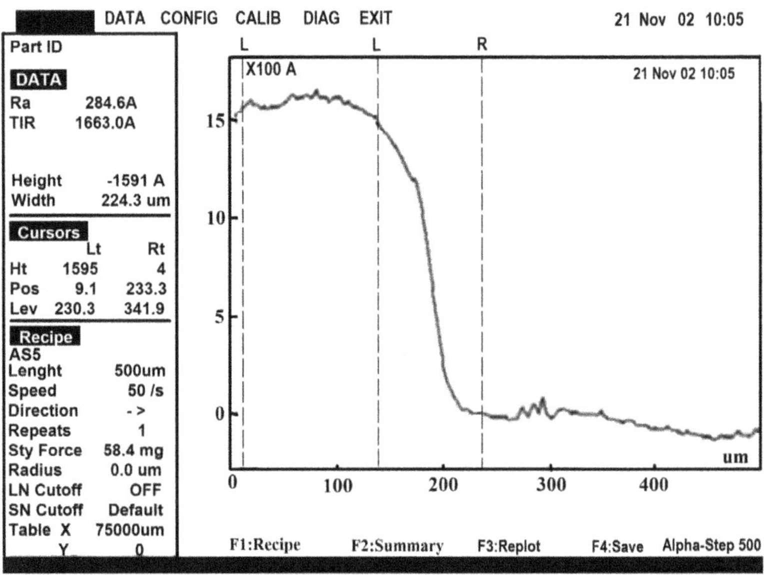

Fig. 5.13 Recorded profilogram of the Ag thin film on the Si substrate.

thickness. Here, a hard stylus moves by the surfaces of the measured sample. If the thin film has a border shaped as a step, the stylus moving horizontally across the sample should show the height of this step. An electro-magnetic sensor belonging to the profilometer detects the vertical deflection of the stylus and records it. This device enables us to measure thin films thickness with high resolution (up to 10 Å). Measured and recorded characteristics appear also as roughness of the films. Figure 5.13 presents the profilogram of a silver thin film deposited with the step on the silicon wafer surface. As shown in Fig. 5.13, the recorded profilogram has casual errors in the measurement due to the substrate roughness. The thickness estimation here depends on the operator's setting that brings additional errors. If we take into account that atom dimensions are 2–4 Å, recorded data is questionable compared to the real value of the measured thickness. Additional deficiencies of this method are as follows: penetration and scratching of the film by the hard stylus, preparation of the specific samples (with steps) for the thickness measurement, high cost of the measuring devices.

Table 5.1 Selected techniques for thin films' thickness measurement.

Method	Range	Accuracy or Precision	Comments
Multiple-beam FET	30–20,000 Å	10–30 Å	A step and reflective coating required
Multiple-Beam FECO	10–20,000 Å	2 Å	A step and reflective coating and spectrometer required; accurate but time-consuming
VAMFO	800 Å–10 μm	0.02–0.05%	For transparent films on reflective substrates; Nondestructive
CARIS	400 Å–2 μm	10 Å–0.1%	For transparent films; Nondestructive
Step gauge	500–15,000 Å	~200 Å	Values for SiO_2 on Si
Ellipsometry	A few Å to a few μm	1 Å	Transparent films; complicated mathematical analysis
Stylus	20 Å to no limit	A few Å to <3%	Step required; simple and rapid
Weight measurement	<1 Å to no limit		Accuracy depends on knowledge of film density
Crystal oscillator	<1 Å to a few μm	<1 Å to a few %	Nonlinear behavior at larger film thicknesses

Table 5.1 represents various selected methods for the thickness measurement. This variety of devices and techniques is due to the plurality of thin films which are used in the world. There are decorative, metallurgical, semiconductor, protective, functional films and coatings applied in microelectronics, optics, photovoltaics, plasmonic devices and other systems. This table presents not only selected methods of measurement but also typical measurement ranges and accuracy levels. Usually, such methods as ellipsometry, atom force microscopy, scanning electron microscopy, field ion beam, Auger spectroscopy and other high-energetic techniques require very complicated and high-cost equipment, which cannot be used in training laboratories devoted for training the undergraduate students. Therefore, in our description, we will consider more simple methods,

such as weight measurement, Michelson interferometry and evaluation of the thin films thickness using the measured transmittance characteristics of the films recorded by help of spectrophotometers.

5.2.2 *Weighting method*

This method consists of measurement of the weight of the substrate before and after thin film deposition. Mass difference divided on the film's surface area and density brings us a value of the thickness, *d*:

$$d = \frac{M_f - M_s}{\rho_f A} \tag{5.20}$$

where M_s is the weight of a pure substrate, M_f is the substrate with the deposited film, ρ_f is the density of the film's material and A is the measured or calculated surface area. The measurement error here is defined by the instrumental error of the balance and errors in determination of the surface area and the film's density, which is lower than in the solid material. Theoretically, the film's density may be calculated using the following formula:

$$\rho_f = \frac{nM_a}{V_c N_A} \tag{5.21}$$

where n is the number of atoms per unit cell, M_a is the atomic mass (g/mole), V_c is the volume of the unit cell (cm^3) and N_A is the Avogadro number ($6.02 \cdot 10^{23}$ $mole^{-1}$). Figure 5.14 represents a balance enabling measurement with precision of μg.

The weighting method is very simple and rough; however, it allows to provide operative measurements. To increase the accuracy of the measurements, several tests (usually 3) should be provided. Averaged result is used. Evidently, this method of thickness measurement allows us to measure its average value.

5.2.3 *Interferometer of Michelson*

Interferometry relates to the optical methods of thin films' characterization. Interferometry is based on the splitting of the light ray on two

Fig. 5.14 Microbalance for weighting the substrates with and without thin films.

coherent rays that interfere after they have merged. The image obtained due to interference of spitting rays using interferometer gives the information about the optical path difference. The known wavelength of the applied light source allows us to measure the thickness of thin films in the units of the wavelength.

When two coherent waves reflected from both surfaces of the transparent thin film or from the surface and substrate of the opaque thin film come to one point with the phase shift, the resulting wave appears as the interference image. Both these cases are shown in Fig. 5.15. Figure 5.15(a) explains how the phase shift is formed while light reflects from both borders of the transparent film. Here, the difference between optical paths for reflected rays R_{12} and R_{23} is shown. In the case of the opaque thin film, difference of optical paths is presented on Fig. 5.15(b) for rays R_{12} and R_{13}.

Two coherent waves reflected from the studied sample may be presented in the following form:

$$R_{12} = E_0 \mathrm{Sin}\omega t \text{ and } R_{23} = E_0 \mathrm{Sin}\left(\omega t + \varphi\right) \tag{5.22}$$

where ω is the angular frequency of the waves and φ is the phase shift of the R_{23} wave that appears due to different optical path of the light. Optical path difference, as shown from Fig. 5.15, is equal to

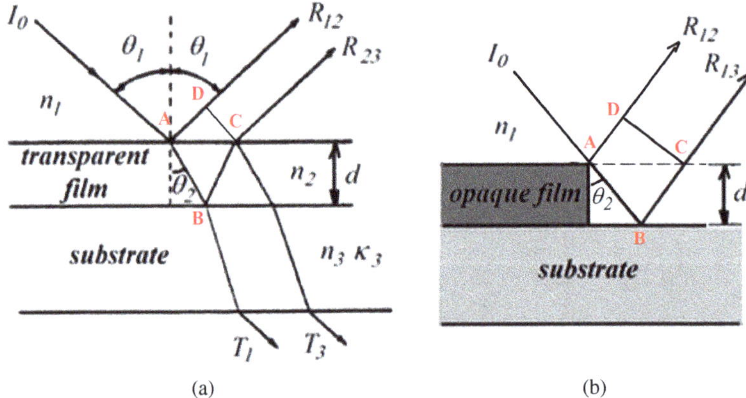

Fig. 5.15 Coherent light waves reflected from different surfaces have different phase shifts due to the different wave path. (a) represents the full image in the case of a transparent thin film and (b) illustrates the reflection in the case of an opaque thin film.

$\Delta L = n_2(AB + BC) - n_1 AD$ for the Fig. 5.15(a) and $\Delta L = n_1(AB + BC) - n_1 AD$ for the Fig. 5.15(b). These relations may be transformed as follows:

$$AB = BC = \frac{d}{\cos\theta_2}; AD = AC \cdot \sin\theta_2; AC = 2d \cdot tg\theta_2;$$

$$\Delta L = \frac{2n_2 d}{\cos\theta}(1-\sin^2\theta_2) \tag{5.23}$$

$$\Delta L = 2dn_2\cos\theta_2 \text{ for transparent films and } \Delta L = 2d\cos\theta_2 \text{ for opaque films} \tag{5.24}$$

The phase shift is equal to:

$$\varphi = \frac{2\pi d}{\lambda}\sin\theta_2 \tag{5.25}$$

And an intensity of the resulting sinusoidal wave (the sum of R_{12} and R_{23}) is given by

$$I = 4I_0\cos^2\frac{\varphi}{2} \tag{5.26}$$

Analysis of Eq. (5.26) shows that one can see the maxima and minima of light and positions of these extrema are defined by the thickness of the thin film. Intensity of resulting light will be maximal when $\cos\frac{\varphi}{2} = 1$. This is right for the following conditions:

$$\frac{\varphi}{2} = m\pi \text{, where } m = 0, 1, 2, \ldots \tag{5.27}$$

Combining Eqs. (5.25) and (5.27), we obtain the intensity maxima

$$2m\pi = \frac{2\pi d}{\lambda}\sin\theta_2 \Rightarrow d\sin\theta_2 = m\lambda, \text{ where } m = 0, 1, 2, \ldots \tag{5.28}$$

And for intensity minima

$$d\sin\theta_2 = \left(m + \frac{1}{2}\right)\lambda \tag{5.29}$$

Presented ideas were realized in the device called Michelson's interferometer. This device enables us to watch the interference image formed by two rays reflected from two mirrors. Figure 5.16 represents two possible principle schemes for the interferometer arrangement.

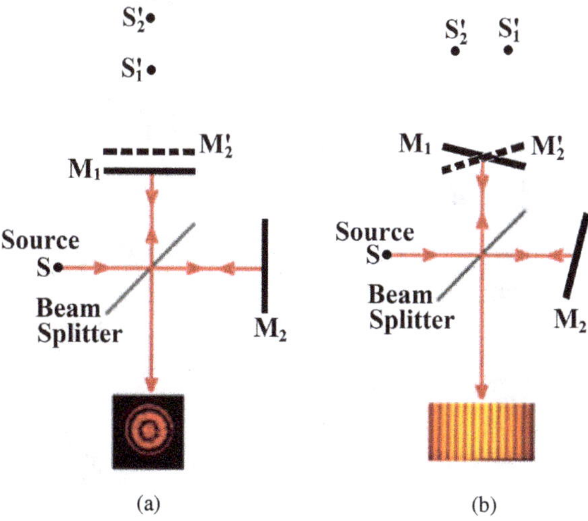

Fig. 5.16 Formation of interference pattern in a Michelson interferometer.

A light ray comes out from a source S and splits on the beam splitter onto two coherent rays which reflect from mirrors 1 and 2 and come together toward the observer. An observer represents a screen or a video recorder. The interference pattern is the result of a combination of these coherent rays. Properties of the interference pattern are defined by the nature of the light source and exact orientation of the mirrors and the beam splitter. Difference between both schemes is in the position of mirrors. Virtual sources S_1' and S_2' are images of the origin source S. Their orientation determines the type of obtained image, concentric rings or parallel straight lines.

Figure 5.17 represents the home-made laboratory interferometer arranged according to the Michelson scheme (a) and the interference pattern recorded on the tantalum thin film with scratch.

To measure the thickness of the thin film, we need to arrange the sample with the thin film on the surface of the mirror M_1, and analyze the recorded pattern. Thickness of the film can be calculated using the following formula:

$$d = \frac{d_1}{\Delta}\frac{\lambda}{2} \tag{5.30}$$

The precision of this measurement method is defined by the accuracy of measurements on the interference pattern and by the

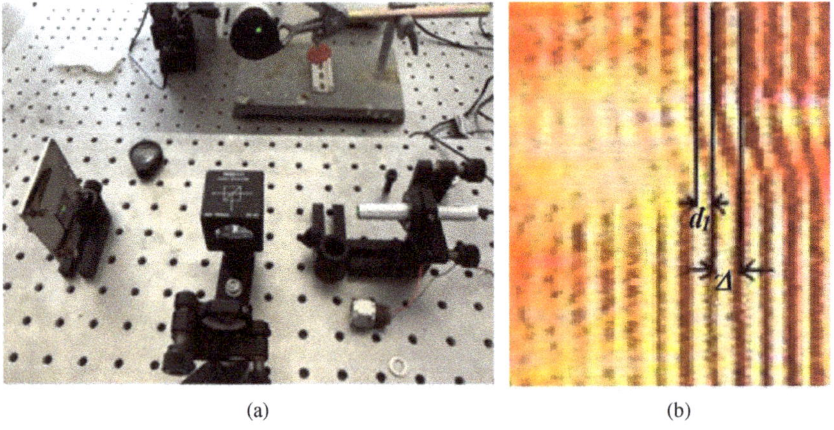

Fig. 5.17 The laboratory interferometer of Michelson (a) and the interference pattern recorded on the scratched tantalum thin film (b).

wavelength of the applied light source, the type of laser and does not exceed ~50 nm. The significant deficiency of the measurement by this method is requirement of an abrupt step or a scratch on the film's surface to obtain a shift of the interference image on the step. At the same time, this method is operative, more accurate than the weighting method and may be successfully used in the training laboratory.

5.2.4 *Estimation of the thickness using the transmittance characteristics*

Figure 5.18 represents the measured transmittance characteristics of the pure glass sample and the same glass coated by a transparent conductive coating. This is a commercial glass coated by the indium-tin oxide (ITO) thin film bought in the store. The main parameters of thin coatings such as the refraction and extinction coefficients and the thickness of the ITO film were the subject of further applications. As can be seen from the transmittance spectrum, it has the interference fringes occurring when the film surface is reflecting without much scattering-absorption in the bulk of the film.

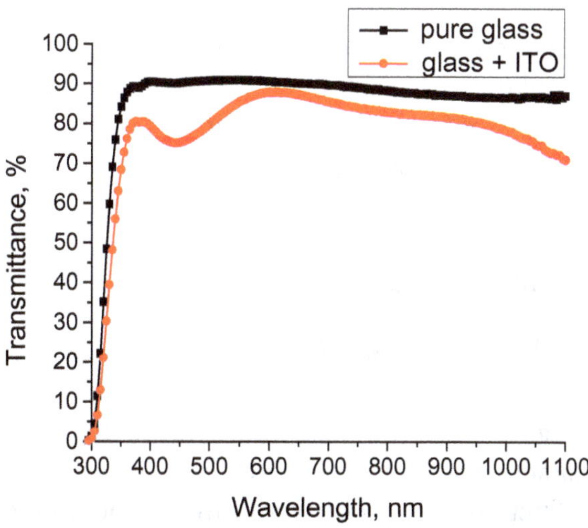

Fig. 5.18 Measured transmittance characteristic of the commercial ITO on glass.

This effect is caused by multiple reflections of light from the interfaces between the thin film, glass substrate and air. The transmittance of a thin film sandwiched between two media (air and glass) at a normal light incidence can be expressed in the following form:

$$T = \frac{T_1 T_2}{1 + R_1 R_2 - 2\sqrt{R_1 R_2}\cos\left(\frac{4\pi}{\lambda} nd\right)} \tag{5.31}$$

where R_1 and R_2 are the intensity reflection coefficients at the air-film and film-glass interfaces, respectively, $T_1 = 1 - R_1$ and $T_2 = 1 - R_2$ are corresponding intensity transmittance coefficients, d is the film thickness and n is the refractive index of the thin film. Using the Fresnel formulae for the reflection and transmission coefficients, Eq. (5.31) may be described as follows:

$$T = \frac{\left[1 - \left(\frac{1-n}{1+n}\right)^2\right]\left[1 - \left(\frac{n - n_{\text{glass}}}{n + n_{\text{glass}}}\right)^2\right]}{1 + \left(\frac{1-n}{1+n}\right)^2 \left(\frac{n - n_{\text{glass}}}{n + n_{\text{glass}}}\right)^2 - 2\left(\frac{1-n}{1+n}\right)\left(\frac{n - n_{\text{glass}}}{n + n_{\text{glass}}}\right)\cos\left(\frac{4\pi}{\lambda} nd\right)} \tag{5.32}$$

where n_{glass} is the refractive index of the glass substrate. This equation includes two unknown variables: Thickness d and refraction coefficient n. Solution here is possible by the iteration method. The maximum and minimum transmittance, according to Eq. (5.32), are defined by the following relations correspondingly:

$$\begin{cases} 2nd = \mathrm{m}\lambda \ (m = 1, 2, 3, \ldots \text{ for a maximum} \\ 2nd = \frac{(2\mathrm{m} + 1)\lambda}{2} (m = 0, 1, 2, \ldots) \text{ for a minimum} \end{cases} \tag{5.33}$$

Now, we can choose the maximum and minimum points in the measured characteristics and calculate the thickness by choice and change the refraction coefficient. Verification should be made by calculation of the transmittance and comparing with the measured values. Calculated thickness of the ITO commercial films in this case was 150 ± 10 nm or ~150 nm.

Such main optical parameters of thin films as a refraction index $n(\lambda)$ and an absorption coefficient $\alpha(\lambda)$ are spectral functions and depend on the wavelength, λ. Therefore, to extract these parameters and thickness from measured spectral characteristics, we need to solve the very difficult task with the number of variable values more than number of equations. These tasks require iteration solving processes or various approximations which bring approximate solutions. Calculation of the thickness by measured transmittance characteristics may be done by various approximate methods: An envelope method or a fitting method or their combination. Detailed consideration of mentioned methods is beyond the scope of this book.

Chapter 6

Electrical properties of thin films

6.1 Introduction

6.1.1 *Conductivity of metals*

If we take a piece of a metal and pass a current through it as shown in Fig. 6.1, one can see that this current will be proportional to the applied voltage (Ohm's law, see Eq. (5.4)).

Figure 6.1 illustrates the electrical current motion in the metal piece. Metal represents a crystalline structure including the electron gas formed due to overlapping of conductive and valence bands in the material. The valency electrons are also the conductive free electrons. Resistance of the metal is due to collisions of electrons with the lattice atoms, impurities and crystal defects. Resistance is directly proportional to the length of the conductive material and inversely proportional to its area. A proportionality coefficient called the resistivity, ρ, is reciprocal to the value called the conductivity, σ.

An electrical current density may be described by the empirical equation of Ohm's law:

$$j = \sigma E \tag{6.1}$$

where E is the electrical field strength. If we assume that each atom of the metal contributes one electron to the conduction band (this is

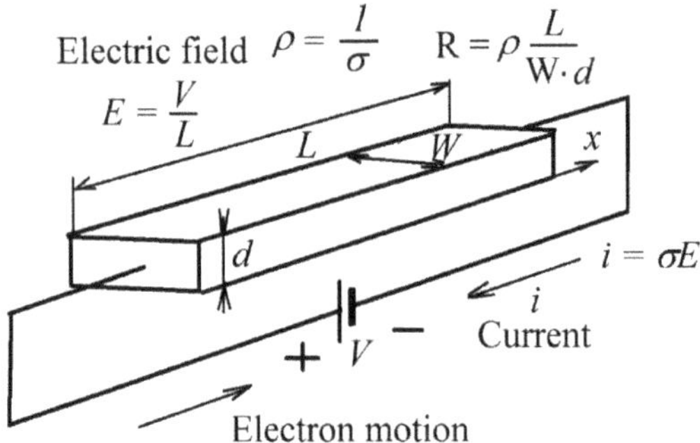

Fig. 6.1 Electrical current motion through the metal piece.

right for such metals as Au, Ag, Cu), that a number of electrons in the unit of volume will be equal to the number of atoms in the same volume:

$$N_0 = \frac{\gamma}{M_a} N_A \tag{6.2}$$

where γ is the density of a metal, M_a is its atomic mass and N_A is the Avogadro number. These electrons move inside the metal randomly. However, if we apply an electrical field, they begin to move opposite to the applied field under a force $F = qE$. So, a drift current appears. Due to collisions of electrons with the lattice, impurities and crystal defects, an opposing force representing the resistance also influences electrons and the equation of electron motion will be as follows:

$$m_e \frac{\partial v}{\partial t} + \delta v = qE \tag{6.3}$$

where v is the drift velocity of electrons, δ is the proportionality coefficient and m_e is the mass of electron. If we assume that the process is in the steady state and $v = v_f =$ const, that the first term in the left part will be zero, we obtain

$$\delta = \frac{qE}{v_f} \tag{6.4}$$

Substituting the relation (6.4) into Eq. (6.3), we obtain the following motion equation:

$$m_e \frac{\partial v}{\partial t} + \frac{qE}{v_f} v = qE \tag{6.5}$$

And the solution of the motion equation is

$$v = v_f \left(1 - e^{\frac{qE}{m_e v_f} t}\right) \quad \text{where } \tau = \frac{m_e v_f}{qE} \text{ is the relaxation time} \tag{6.6}$$

The relaxation time defines the average time between two consequent collisions. Therefore, one can define $\lambda = v\tau$ as a mean free path of the electron. Now, we can find the value of v_f:

$$v_f = \frac{q\tau}{m_e} E = \mu E \text{ where } \mu \equiv \frac{q\tau}{m_e} \text{ is the mobility of electrons} \tag{6.7}$$

A current density is proportional to the drift velocity:

$$j = qN_0 v_f = \sigma E \tag{6.8}$$

From this equation, we can find the conductivity:

$$\sigma = \frac{qN_0 v_f}{E} = \frac{q^2 \tau N_0 E}{m_e E} = \frac{q^2 \tau N_0}{m_e} = q\mu N_0 \tag{6.9}$$

We showed here a simplified theory of conductivity. It has several limitations, however, the formula (6.9) describing the conductivity of materials is as relevant for metals as for semiconductors. Obviously, the electron mass m_e means the effective mass, and not the mass of a free electron due to movement inside a material, metal or semiconductor when a charged carrier is exposed to a crystal field.

In general, all materials may be classified according to their conductivity. Figure 6.2 presents such a classification. As shown in this picture, all materials are divided into three big groups: metals, semiconductors and insulators.

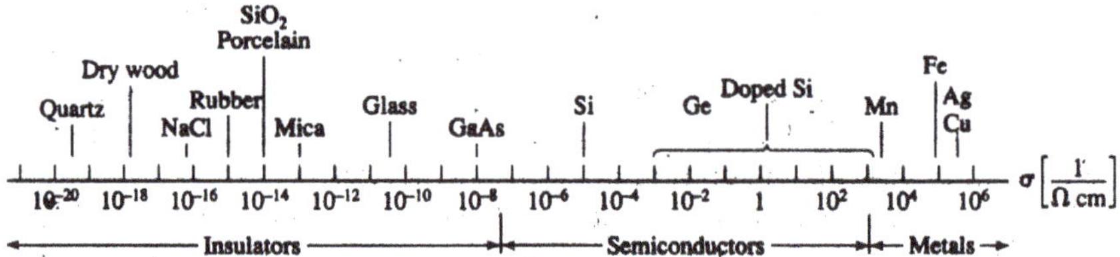

Fig. 6.2 Classification of materials by their conductivity.

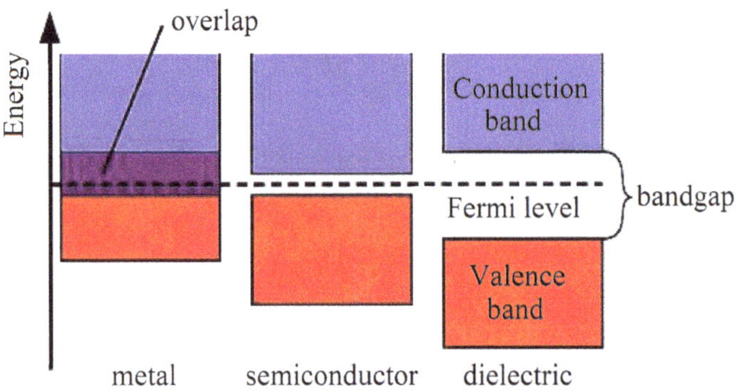

Fig. 6.3 Energetic structure of different materials defined by their type of bonding.

This classification is defined basically by the type of bonding inside the materials. Figure 6.3 illustrates the basic difference between various materials from the point of view of the electron structure and the type of bonding of considered materials.

In the metals, due to overlapping of the conduction and valence bands, all free electrons are shared and each electron belongs to each atom inside the metal piece. In the ion-type materials, there are no free electrons. These materials are insulators and a direct current conductivity in these materials may be realized only by defects and impurities. In the materials with covalent bonding, such as most semiconductors and dielectrics, the conductivity depends on the number of excited electrons at given temperature and on the bandgap width. According to this definition, the transparent conductive oxide thin films built from indium-tin oxide or zinc oxide alloyed by aluminum, for example, belong to the wide-bandgap semiconductors, approximately up to 3.5–3.8 eV.

Evidently, to measure the conductivity or a reciprocal value, i.e., the resistivity of the bulk materials, we can realize the measurement scheme according to Fig. 6.1. However, in the case of two-dimensional materials, thin films, this method is not applicable. Especially when it concerns the very thin non-continuous, islet films.

Firstly, let us consider the thin solid metal films. According to the Matthiessen rule, the resistivity of the thin metal film may be presented as a simple arithmetic sum of the individual contribution of collisions of electrons with the lattice, impurities and defects:

$$\rho = \rho_{\text{th}} + \rho_{\text{imp}} + \rho_{\text{def}} \tag{6.10}$$

where the first summand ρ_{th} only depends on the temperature. The number of collisions with phonons increases with the temperature growth according to the following known relation:

$$\rho_2 = \rho_1[1 + \alpha(T_2 - T_1)] \tag{6.11}$$

where α is the temperature coefficient of resistance and $(T_2 - T_1)$ is the temperature difference. Second and third summands in Eq. (6.10) are not dependent on temperature. So, the resistivity dependence on temperature in thin continuous metal films may be approximately illustrated by a linear behavior as shown in Fig. 6.4.

In the case of thin films, additional term responsible for collisions of moving electrons with the film boundaries should be attached to Matthiessen's equation and for the film's resistivity ρ_f:

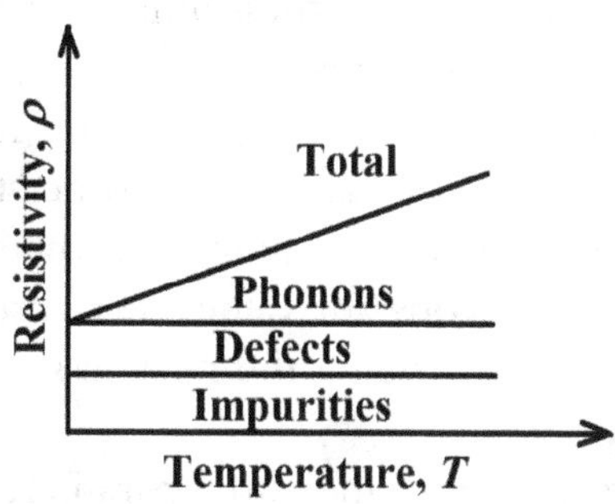

Fig. 6.4 Resistivity behavior according to Matthiessen's rule.

$$\rho_f = \rho_{\text{th}} + \rho_{\text{imp}} + \rho_{\text{def}} + \rho_b \tag{6.12}$$

where ρ_b is the nonlinear term dependent on the electron scattering on the film's boundaries. If we denote a thickness factor $\gamma = d/\lambda$, as described in Chapter 1, an influence of the thickness on the resistivity in the case $\gamma \ll 1$ may be described by the following approximate relation:

$$\rho_f = \rho \frac{1}{\frac{3\gamma}{4}\left(ln\frac{1}{\gamma} + 0.423\right)} \tag{6.13}$$

This relation may be obtained by the analytic way during analysis of the known Boltzmann transport equation (BTE) for electrons in the simplified form:

$$-\frac{q}{m_e}\left(E + \frac{1}{c}v \times H\right)\text{grad}_v f + v\,\text{grad}_r f = -\frac{f - f_0}{\tau} \tag{6.14}$$

where f is a non-equilibrium electronic distribution function, f_0 is the electronic distribution at equilibrium, v is the electron velocity and τ is the relaxation time for return to equilibrium. In the case of thin metal continuous films and in the absence of a magnetic field and a temperature difference at the film area, after integration we obtain the relation (6.13).

From a practical point of view, the resistivity of very thin films significantly increases and the thickness influences on the other properties. For example, Fig. 6.5 presents the properties of the thin molybdenum films deposited by the sputtering method. Here, the thickness of the fattest was of 122 nm. All other films shown in the plots as the measured points were thinner. In the sputtering process, the sputtering rate is the constant value, therefore thicknesses of all films may be calculated by dividing the thickness of the fattest film on the processes' durations.

As shown in Fig. 6.5, both measured parameters rise sharply with the film's thickness decreasing.

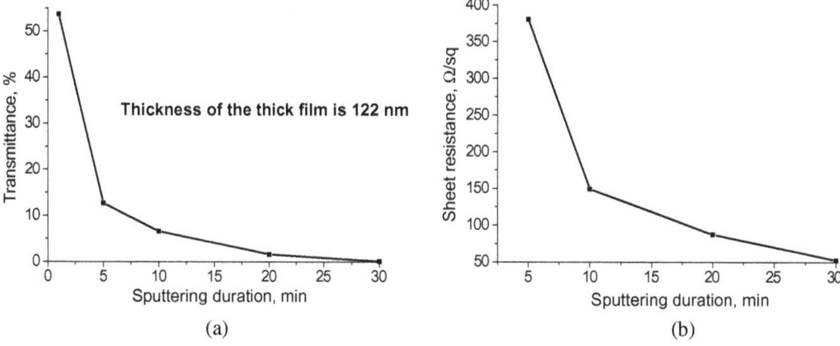

Fig. 6.5 Properties of thin molybdenum films deposited by sputtering: (a) transmittance dependence on the sputtering duration; (b) sheet resistance dependence.

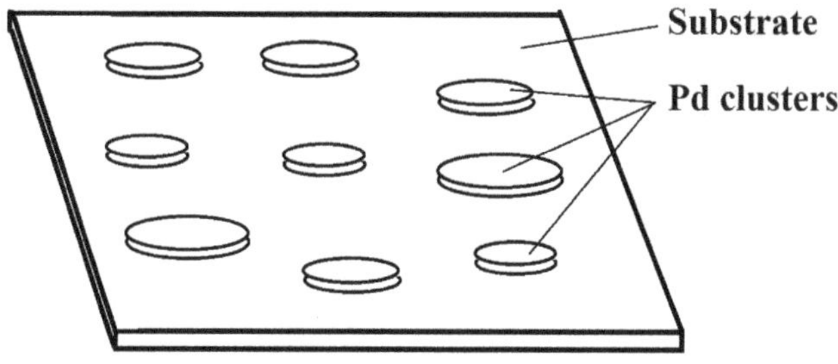

Fig. 6.6 Palladium islets thin film deposited on the quartz substrate by sputtering.

When we intend to prepare an alloy by the sputtering or evaporation method, the composition of the grown thin films is analyzed as per the Rule of Nordheim. Especially for the binary alloy with components *A* and *B*, this Rule looks as follows:

$$\rho = X_A\rho_A + X_B\rho_B + CX_AX_B \tag{6.15}$$

where X_A and X_B are atomic concentrations of the components and C is the material constant.

Island metal thin films also can provide an electrical current, however in such a case, the resistivity is dependent on the shape of the metal islets and the distance between them. Figure 6.6 represents a quartz substrate with deposition of sputtering discs from palladium.

The palladium thin films were used to prepare the conductive and transparent in the near ultraviolet range electrodes for application in the transmission television tubes. These tubes were used before invention of the charge-coupled devices (CCD). As known, the usual transparent conductive coatings cannot be used in the near-ultraviolet range due to the red border situated near 300 nm. At the same time, the thin island metal coatings can provide the satisfactory properties. Figure 6.7 presents the dependencies of palladium thin island films properties on the annealing temperature. Annealing was provided in vacuum after the deposition.

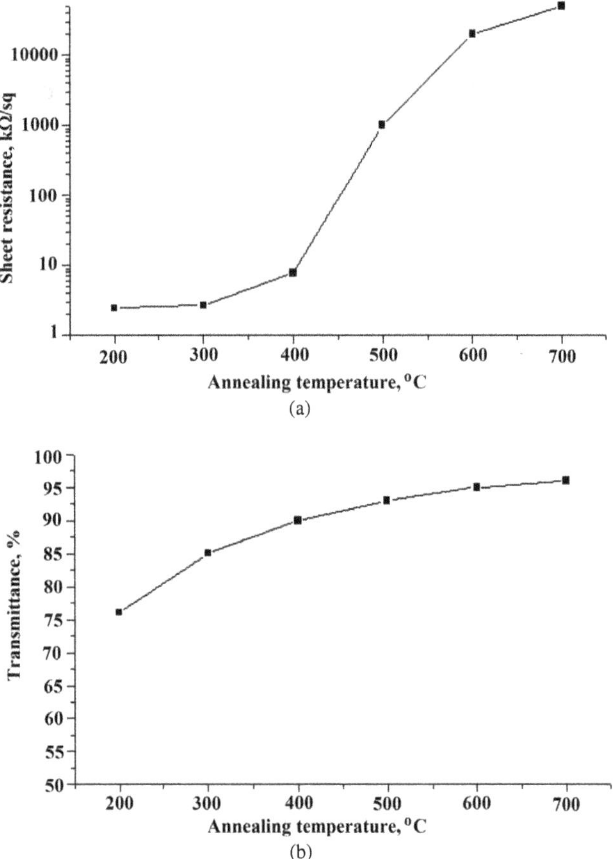

Fig. 6.7 Properties of thin Pd films under the post-deposition annealing: (a) sheet resistance of the Pd thin island film; (b) transmittance of the thin Pd film.

As shown in Fig. 6.7, the surface resistance of the layers increases sharply under annealing to several orders of magnitude (Fig. 6.7(a)) and at the same time, the transmittance of the films increases by 20% approximately (Fig. 6.7(b)). This indicates that these films may be annealed up to 300°C without dramatic changes in their properties. After that, changes begin in the shape of the islets from disc form to the half-sphere or sphere and increase in the distance between metal islets. In the case of the non-continuous thin film, the resistivity will be sum of two summands: first one represents the metal resistivity described by Eq. (6.12) and the second one will be defined by the jumping between neighboring islands. The jumping conductivity may be realized by one of three following mechanisms: (1) a quantum tunneling process; (2) a thermionic emission of the electrons from islands under applied electrical field; (conductivity by the surface contaminations. So, the conductivity of the non-continuous thin film may be described by the following equation:

$$\sigma = \sigma_M + \sigma_D = q\mu_n N_0 + q\mu_D N_0 e^{\frac{E_A}{kT}} \tag{6.16}$$

where E_A is an energy of the transfer activation depending on the diameter of the islands and the distance between them, μ_n is the mobility of the electrons dependent on the three above-mentioned mechanisms. Thus, the surface coated by thin metal islands acquires novel semiconductor properties, which may produce some additional effects, for example the plasmon resonance. Such materials with changed properties are called meta-materials.

6.1.2 *Conductivity of dielectric materials*

An ideal dielectric material has no free electrons or ions. Therefore, in the absence of an external electrical field, this material is electroneutral. However, under influence of the external field, all molecules (usually, these molecules are polar) set with relation to the acting field. This process is called the polarization process. Each molecule or atom creates an electrical dipole momentum p measured by $p = \left[\frac{C}{m^2}\right]$.

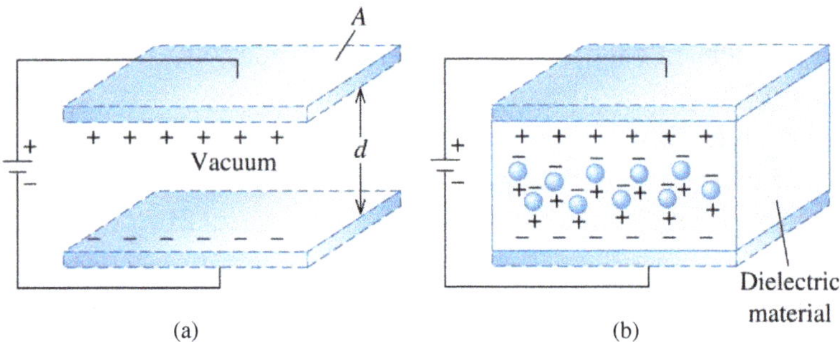

Fig. 6.8 A flat capacitor system that is empty (a) and fully filled by a dielectric material (b).

Momentums of all molecules in the material are summarized and produce polarization $P = \Sigma p_i$. Figure 6.8 presents a flat capacitor system that is empty (a) and fully filled by a dielectric material (b).

Under an applied electrical field, the internal volume of the capacitor polarizes. The capacity of the aerial (vacuum) capacitor (Fig. 6.8(a)) is defined by the capacitor geometry (an area, *A* and *a* distance between flat electrodes, *d*) and the permittivity of vacuum, ε_0:

$$C = \frac{\varepsilon_0 A}{d} \tag{6.17}$$

If we designate the applied electrical field by $E = V/d$, where V is the applied voltage, the capacitor will store an energy according to the following equation:

$$W = \frac{1}{2} C V^2 \tag{6.18}$$

A dielectric material inserted in the capacitor (Fig. 6.8(b)) changes the stored charge in the value ε_r which is called the relative permittivity or a dielectric constant:

$$\frac{C_1}{C} = \frac{Q_1}{Q} = \varepsilon_r \tag{6.19}$$

As is known, the permittivity is the ability of a material to polarize and store a charge within it. An additional charge stored in the fully filled capacitor is proportional to the polarization:

$$PA = Q_1 - Q = C_1 V - CV = CV(\varepsilon_r - 1) \rightarrow P$$
$$= \varepsilon_0 E(\varepsilon_r - 1) \rightarrow \varepsilon_r = 1 + \frac{P}{\varepsilon_0 E} \quad (6.20)$$

If we designate the relation $\chi = \frac{P}{\varepsilon_0 E}$ as the dielectric susceptibility and $\alpha_e = \varepsilon_0 (\varepsilon_r - 1)$ as the electronic polarizability, then:

$$\varepsilon_r = 1 + \chi; \; P = \alpha_e E; \; D = \varepsilon E = \varepsilon_0 \varepsilon_r E = \varepsilon_{0E} + P \quad (6.21)$$

Dielectric material with purely covalent bonding, such as diamond, has clouds of negatively charged electrons symmetrically distributed about the positively charged nucleus. An applied electric field pushes the electrons in the direction opposite the field and the nucleus toward the field. The atom distorts, and the centers of positive and negative charges are not coinciding now. So, each atom has now an electric dipole moment $p = q\delta$. For N atoms per m^3, the total electric dipole moment $P_S = Np = Nq\delta$. Figure 6.9 illustrates this influence of the applied electrical field on the covalent bonded system.

The covalent symmetric molecules produce a dipole momentum under electrical field as shown in Fig. 6.9(a). The length of the bonds depends on the applied field value, see Fig. 6.9(b).

Ion crystals, for example NaCl, contribute additionally to the polarization from orientation of ions. However, the polarization of ions requires additional time to obtain the direction of the external electrical field. Materials with the asymmetric crystal structure have a natural polarization without application of external electrical fields. Such materials as $BaTiO_3$ or ZnO show a spontaneous ionic

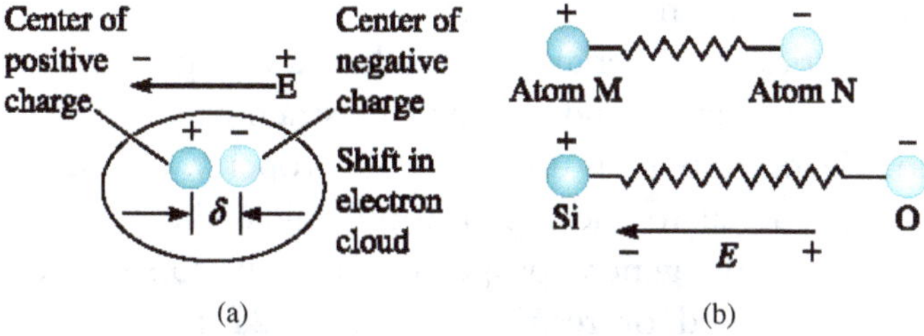

Fig. 6.9 Illustration of the influence on the polarization of the covalently bonded system (a) and (b) the influence of the applied electrical field on the bond's length.

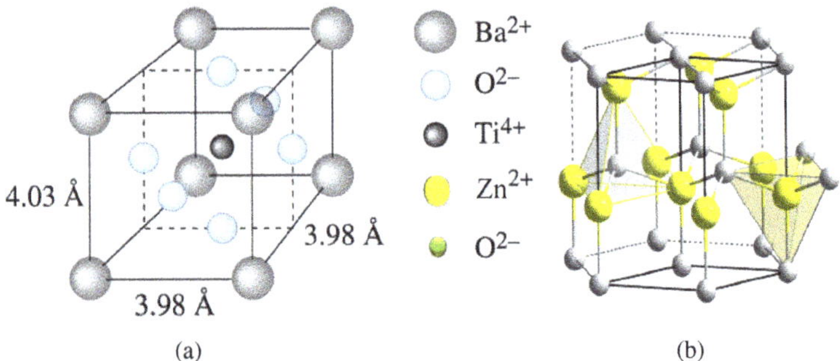

Fig. 6.10 Asymmetric crystal structures: (a) BaTiO3 and (b) ZnO.

polarization and have a dielectric susceptibility of several thousands. Their crystal structures have been presented in Fig. 6.10.

Under external influence, these materials show several nonlinear electrical properties associated with their spontaneous polarization. These properties relate mechanical and electrical properties and are called electrostriction and piezoelectricity, pyroelectricity and ferroelectricity. All dielectric materials have an electrostriction. This property consists of the small movement of ions in the crystal lattice when exposed to an external electric field. Positively charged ions are pushed out in the direction of the field, while negatively charged ions are pushed out in the opposite direction. This displacement gradually accumulates in the entire thickness of the material and leads to a general deformation or elongation of the body in the direction of the field. The thickness of the substance decreases in orthogonal directions. The strain or the electrostriction produced by the application of the electric field is proportional to the square of the electric polarization. So, reversing the electric field will not change the strain.

Piezoelectrics are materials that produce a voltage difference upon the application of a stress and produce the strain when an electric field is applied. Piezoelectricity is the reversible property. Piezoelectricity is linearly proportional to the polarization unlike the electrostriction. Electrostriction is the general property of all insulators whereas piezoelectricity is confined or restricted only to 21 non-centric crystal classes. The ability of a material to spontaneously polarize and produce a voltage due to changes in temperature is called pyroelectricity.

And all materials showing spontaneous and reversible dielectric polarization are called ferroelectrics.

In AC circuits, a dielectric material always operates in transition conditions. If the dielectric immediately responds to the external electric field change, for ideal dielectric and alternating external field $E = E_0 e^{-i\omega t}$, the polarization will be related with the electrical field as follows:

$$P = \varepsilon_0(\varepsilon_r - 1)\, E_0 e^{-i\omega t} \tag{6.22}$$

A change in surface charge density P requires charge flow of

$$j = \frac{dP}{dt} = -i\omega\varepsilon_0\,(\varepsilon_r - 1)\, E_0 e^{-i\omega t} \tag{6.23}$$

Thus, the conductivity of the dielectric materials under an external alternating electric field will be as follows:

$$\sigma = \frac{j}{E} = -i\omega\varepsilon_0\,(\varepsilon_r - 1) \tag{6.24}$$

This conductivity is the reactive parameter which grows with the frequency increasing and is equal to zero for the DC current. An empirical formula known as the Clausius–Mossotti equation relates the dielectric permittivity of the non-polar dielectric with its electronic polarizability:

$$\frac{(\varepsilon_r - 1)}{(\varepsilon_r + 2)} = \frac{N\alpha_e}{3\varepsilon_0} \tag{6.25}$$

6.1.3 *Conductivity of semiconductors*

As has been mentioned earlier, the conductivity of semiconductor materials depends on the bandgap width and the number of excited charge carriers at a given ambient temperature. Electrons in the conducting band are stimulated due to heat and provide conductivity of matter. Places of electrons that have left a lattice and moved to the conduction band behave like anti-electrons and are called holes.

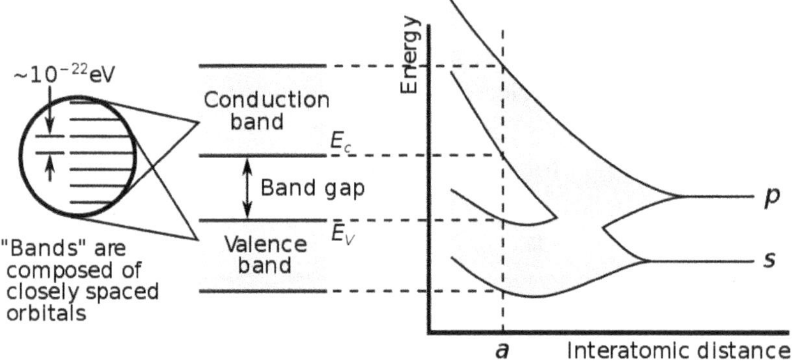

Fig. 6.11 Formation of the energy bands in a semiconductor in the process of approaching atoms constituting the solid crystalline semiconductor.

When atoms of a semiconductor, especially silicon, approach to create a solid crystal structure, the energetic levels of individual atoms split according to the Pauli exclusion principle and form energetic bands, as illustrated in Fig. 6.11. As such an atom's concentration in semiconductors is close to ~10^{22} cm^{-3}, each energy level in the forming energy band also splits on the same number of closely spaced orbitals. Respectively, the distance between these orbitals is equal to ~10^{-22} eV in the energy coordinate system.

If we consider the spectral formula of silicon, ($1s^2\ 2s^2\ 2p^6\ 3s^2\ 3p^2$), the valence band and the conduction band consist of mixed *s* and *p* states. Each energy band in silicon contains 1 s-electron and 3 p-electrons. So, the valence band at ambient temperature equal to absolute zero can contain up to $4N_{Si}$ electrons. With growth of ambient temperature, a part of the electrons will be excited and released and move in the conduction band, forming holes in the valence band. An electron's concentration in the conduction band will be equal to the product of the density of free places in the conduction band and the probability of these free places being captured by electrons:

$$n = 2\left(\frac{2\pi m_e^*}{h^2}kT\right)^{\frac{3}{2}} e^{-\frac{E_c - E_F}{kT}} = N_c e^{-\frac{E_c - E_F}{kT}};N_c \equiv 2\left(\frac{2\pi m_e^*}{h^2}kT\right)^{\frac{3}{2}} \quad (6.26)$$

where E_F designates the Fermi level, E_C designates the bottom line of the conduction band and m_e^* is the effective mass of the electron. Similarly, the concentration of holes in the valence band will be equal to the product of the density of free states in the valence band and the probability of electrons leaving these states:

$$p = 2\left(\frac{2\pi m_h^*}{h^2}kT\right)^{\frac{3}{2}} e^{-\frac{E_F - E_V}{kT}} = N_v e^{-\frac{E_F - E_V}{kT}}; N_v \equiv 2\left(\frac{2\pi m_h^*}{h^2}kT\right)^{\frac{3}{2}} \quad (6.27)$$

where E_V designates the upper bound of the valence band and m_e^* is the effective mass of the hole. Product of p and n from Eqs. (6.26) and (6.27) is always constant and referred to as the mass action law:

$$p \cdot n = N_c N_v e^{-\frac{E_g}{kT}} = n_i^2 \quad (6.28)$$

Concentrations of electrons in the conduction band and holes in the valence band are equal in the pure ideal crystalline semiconductor. So, these concentrations are called the intrinsic concentrations and can be found from the following equation:

$$n_i = N_c e^{\frac{E_i - E_c}{kT}} = p_i = N_v e^{\frac{E_v - E_i}{kT}} = \sqrt{N_c N_v e^{-\frac{E_g}{2kT}}} \quad (6.29)$$

The same equation may be presented as a function of ambient temperature:

$$n_i = 2\left(\frac{2\pi kT}{h^2}\right)^{\frac{3}{2}} \left(m_h^* m_e^*\right)^{\frac{3}{4}} e^{-\frac{E_g}{2kT}} = AT^{\frac{3}{2}} e^{-\frac{E_g}{2kT}} \quad (6.30)$$

This dependence is shown very well in Fig. 6.12.

As shown in Fig. 6.12, the intrinsic concentration dependence on the temperature represents the direct lines in the half-logarithmic diagram. Therefore, the exponential dependence plays a major role in the intrinsic concentration behavior.

To control electrical properties of the semiconductors, they may be doped by materials providing a suitable conductivity type, p or n.

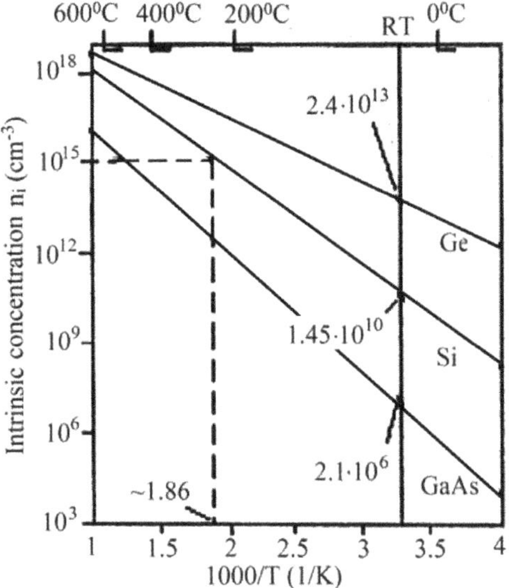

Fig. 6.12 Intrinsic concentration dependence on temperature for chosen semiconductors.

These doping materials providing semiconductors of n-type called donors belong to materials with excess electrons in the external shell. We designate the donor concentration by N_D. Doping materials providing p-type semiconductors, called acceptors, are designated by N_A. After the doping, the semiconductor continues to be electroneutral. According to the charge conservation law, the following equation may be written:

$$(p - N_A) + (N_D - n) = 0 \tag{6.31}$$

where p and n are effective charge carrier concentrations, holes and electrons. Usually, each part of the semiconductor is doped by one of the doping materials, N_A or N_D, which significantly exceeds the intrinsic concentration. All impurities at room temperature are ionized, therefore the charge carrier concentration will be defined by the level of doping.

As such $N_A \gg p$ in the semiconductor doped by acceptors, Eq. (6.31) will be changed:

$$p = N_A - N_D \text{ and } n = \frac{n_i^2}{N_A - N_D} \tag{6.32}$$

Similarly, $N_D \gg n$ in the semiconductor doped by donors, therefore:

$$n = N_D - N_A \text{ and } p = \frac{n_i^2}{N_D - N_A} \tag{6.33}$$

Evidently, the major charge carrier concentration defines the Fermi level value (see Eqs. (6.26) and (6.27)). Thus, one can build the graphic representation of dependence of the Fermi level value on the charge carrier concentration. For example, for the donor-doped silicon, Eq. (6.31) may be written as follows:

$$\left(\frac{n_i^2}{n} - N_A\right) + (N_D - n) = 0 \text{ or } n^2 - n(N_D - N_A) - n_i^2 = 0 \tag{6.34}$$

with the solution in the following form:

$$n = \frac{N_D - N_A}{2} + \sqrt{\left(\frac{N_D - N_A}{2}\right) + n_i} \tag{6.35}$$

And similarly, for the acceptor-doped silicon:

$$p = \frac{N_A - N_D}{2} + \sqrt{\left(\frac{N_A - N_D}{2}\right) + n_i} \tag{6.36}$$

Both Eqs. (6.35) and (6.36) are presented in the graphic form in Fig. 6.13. The presented curves show position of the Fermi level in the silicon as a function of the impurities' concentration. The regions with the value of $3kT$ near the valence and conduction bands show the areas with very high impurities concentrations when the semiconductor behaves as a degenerated semiconductor of high conductivity.

Concentration of charge carriers in the doped semiconductor significantly depends on the ambient temperature. This is due to the

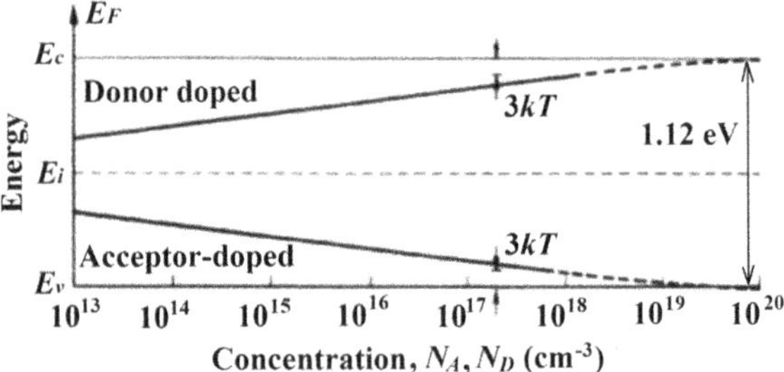

Fig. 6.13 Dependence of the Fermi level in Si on the impurities concentration.

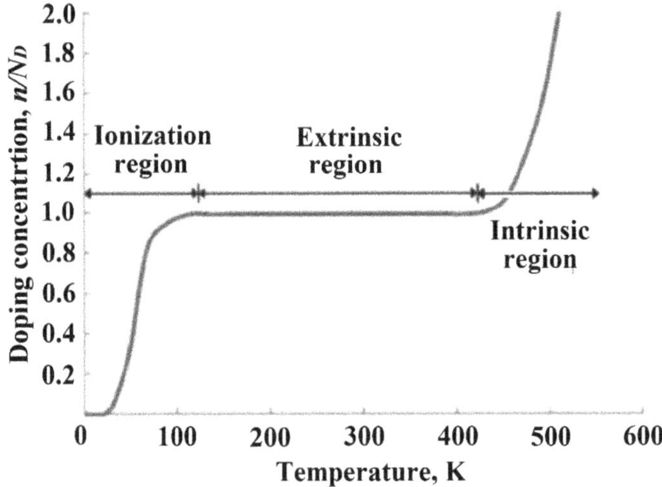

Fig. 6.14 Doping concentration dependence on temperature.

behavior of interstitial impurities in the semiconductor. Figure 6.14 illustrates the donor's behavior in the silicon of n-type.

At low temperatures, all impurities are in the crystal lattice and are not ionized. With temperature growth, the donor impurities ionize and their excess electrons transfer from the valence band to the conduction band, thus all impurities will be ionized at room temperature. Under additional temperature growth, the intrinsic concentration begins to grow according to Eq. (6.30) and the intrinsic concentration exceeds the donor concentration. The semiconductor behaves as the intrinsic material, which limits the use of semiconductor devices.

The second important parameter defining the conductivity of the semiconductor is the mobility. The mobility of charge carriers characterizes the scattering processes through the average scattering relaxation time, $\langle\tau\rangle$:

$$\mu \cong \frac{q\langle\tau\rangle}{m^*} \tag{6.37}$$

where m^* represents an effective mass of electrons. All scattering processes occur simultaneously, therefore, according to the Matthiessen rule, the mobility may be presented by two summands representing the lattice, μ_L, and impurity, μ_i, scattering:

$$\frac{1}{\mu} = \frac{1}{\mu_L} + \frac{1}{\mu_i} \tag{6.38}$$

The expressions for these parameters may be presented in the following form:

$$\mu_L \propto \frac{4q}{m^* \sqrt{9\pi k_B}} T^{-1.5} \tag{6.39}$$

$$\mu_i \propto \frac{8qk_B^{1.5}}{N_i m^* \sqrt{\pi}} T^{1.5} \tag{6.40}$$

where N_i is the concentration of ionized impurities. Figure 6.15 illustrates the mobility dependence on the ambient temperature and the impurities concentration in the silicon.

Combination of two major parameters, charge carrier concentration and mobility, defines the conductivity of a semiconductor. However, a semiconductor contains both types of charge carriers, electrons and holes. Therefore, using relations (6.8) and (6.9), one can write for the current density through the semiconductor the following relation:

$$j = j_n + j_p = \sigma E = q(n\mu_n + p\mu_p)E \tag{6.41}$$

In the case of the intrinsic silicon with $m_n^* = m_p^* = m_0$, the conductivity looks as follows:

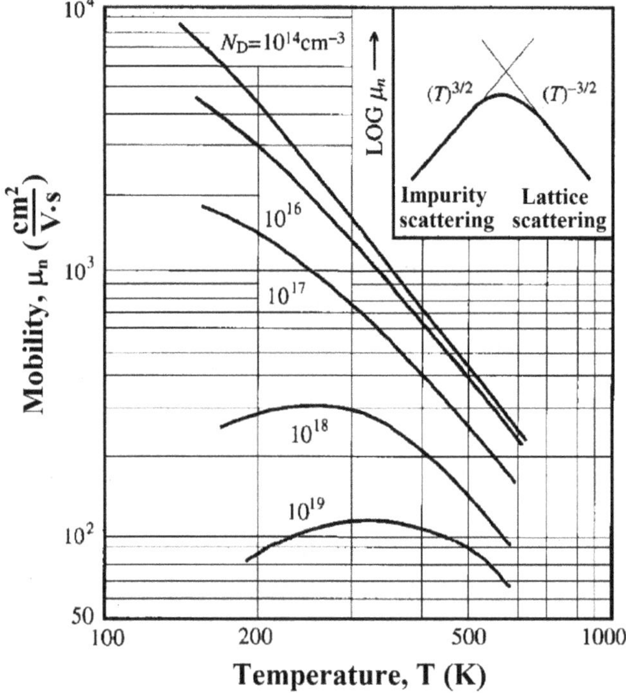

Fig. 6.15 Mobility dependence on temperature for various donor concentrations in Si.

$$\sigma = 4.84 \cdot 10^{15} \left(\frac{m_*}{m_0} \right)^{\frac{3}{2}} T^{\frac{3}{2}} \left(\mu_n + \mu_p \right) e^{-\frac{E_g}{2kT}} \tag{6.42}$$

Evidently, when one of the summands in Eq. (6.41) is low, it may be removed. As both parameters, charge carriers concentration and mobility, are dependent on the ambient temperature, the conductivity, representing the product of these parameters, also depends on temperature as shown in Fig. 6.16. This figure illustrates the influence of the internal parameters of conductivity on the electrical properties of the silicon of n-type.

6.2 Sheet resistance

The resistance of the metal piece is proportional to its resistivity, as shown in Fig. 6.1:

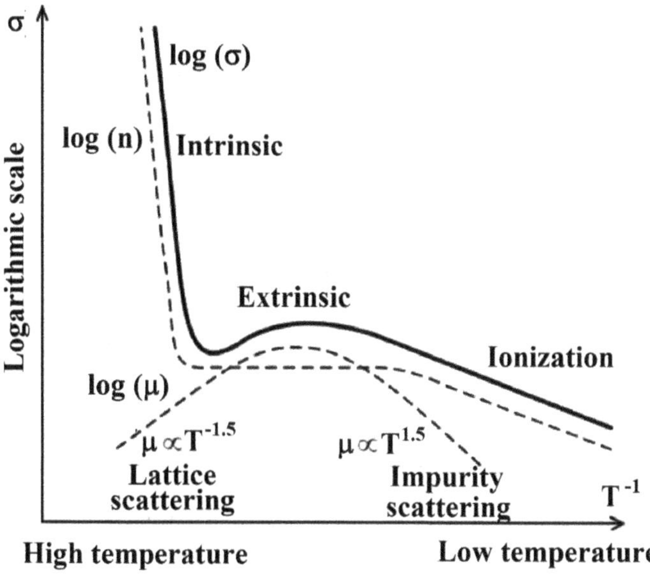

Fig. 6.16 Conductivity dependence on temperature and internal parameters for the Si of n-type.

$$R = \rho \frac{l}{W \cdot d} \tag{6.43}$$

where l is a length, W is a width and d is a thickness of the piece. If we take $W = l$, i.e., the shape of our piece is square, Eq. (6.43) transforms to the following form:

$$R = \frac{\rho}{d} \equiv R_{\text{sq}} = R_{\#} = [\Omega/\text{sq.}] \tag{6.44}$$

here, the value R_{sq} is the surface or sheet resistance. Evidently, all square pieces built from the same material will have the same sheet resistance, regardless of the square dimensions. Sometimes, we cannot provide the suitable conditions for resistivity measurement, as shown in Fig. 6.1. This may be so in the case of a big piece of metal or semiconductor or in the case of thin films. In these cases, electrical characterization of the material may be done using measurement of a sheet resistance by the four-point probe method. Figure 6.17 represents the principle electric circuit for measurement of the sheet resistance.

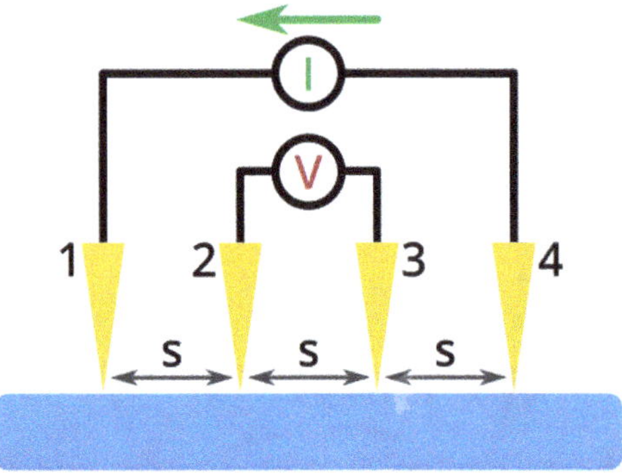

Fig. 6.17 A principle electrical circuit for measurement of the sheet resistance by four-point method.

The measuring system consists of four probes, needles prepared from the tungsten carbide with 0.4 mm diameter are situated linearly at the distance $S = 1$ mm, equipped with individual springs for each needle. To measure the sheet resistance of the sample, it needs to pass a current from the precision current supply through two external needles pinned to the sample surface and to measure the voltage drop between two internal needles using the sensitive voltmeter or galvanometer. The sheet resistance should be calculated using the following relation:

$$R_s = K\frac{V}{I} \tag{6.45}$$

where V is the measured voltage, I is the current passed through external needles and K is the shape's coefficient depending on the thickness of the sample and the distance between needles:

$$K = \frac{\pi}{\ln 2} \approx 4.53 \text{ when } S \gg d \text{ (thin films case)} \tag{6.46}$$

$$\text{And } \rho = 2\pi S\frac{V}{I} \text{ when } S \le d \text{ (the bulk samples case)} \tag{6.47}$$

Fig. 6.18 Typical four-point measuring system suitable for use at high/low temperatures.

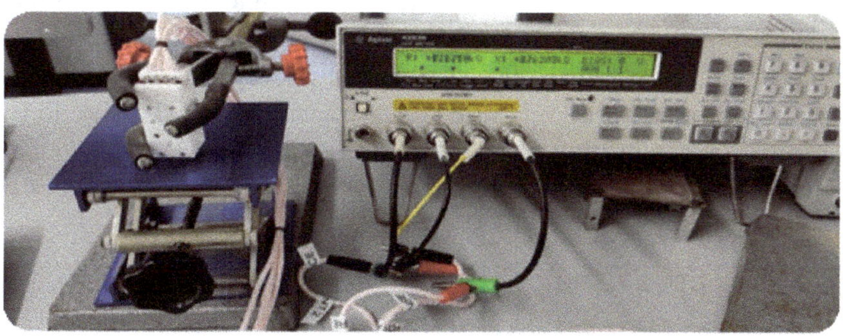

Fig. 6.19 Practical measurement of the sheet resistance.

Figure 6.18 illustrates for example the real measuring system. Sometimes, instead of required current supply and voltmeter, a universal device such as the LCR meter may be applied for measurement of the sheet resistance of thin films. This measurement is illustrated in Fig. 6.19. Here, the sample should be placed on the little table enabling peening the needles of the measuring system to the sample surface. After that the measurement can be provided. It is desirable to provide no less than three measurements in different points of the sample surface and use the averaged measurement's results.

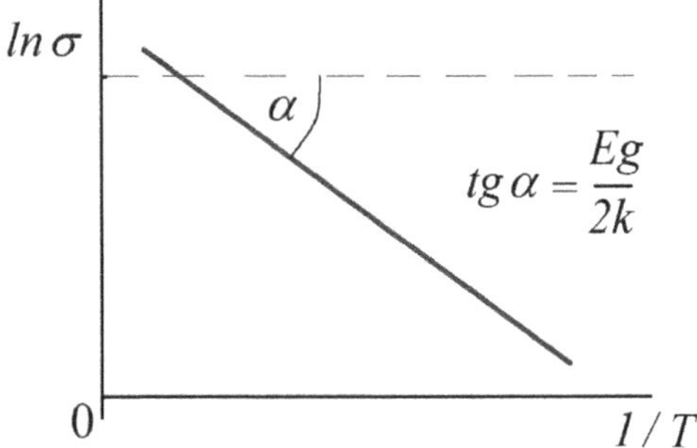

Fig. 6.20 Graphical representation of the semiconductor thin film bandgap dependence on temperature.

Measurement of the sheet resistance of a semiconductor thin film at different temperatures with use of measured thickness enables calculating the bandgap of the film. To provide measurements at different temperatures, the little table, shown in Fig. 6.19, should be changed on the heating plate. According to Eq. (6.42), the dependence of the conductivity on temperature may be written as follows:

$$\sigma(T) = 4.84 \cdot 10^{15} \left(\frac{m_*}{m_0} \right)^{1.5} T^{1.5} \left(\mu_n + \mu_p \right) = \sigma_0 e^{-\frac{E_g}{2kT}} \tag{6.48}$$

A plot of Eq. (6.48) may be presented in the graphical form, as shown in Fig. 6.20.

The bandgap of a semiconductor thin film can be calculated from the slope of the logarithm of the conductivity as a function of temperature.

6.3 Volt-Ampere characterization

Resistors, semiconductors and semiconductor junctions may be characterized by the functional relation between the quantity of the applied voltage and the quantity of the current flowing through the measured sample. These relations are known as the Volt–Ampere (*V–I*) or Ampere–Volt (*I–V*) characteristics. There are many types of *V–I* characteristics. Two types are shown in Fig. 6.21.

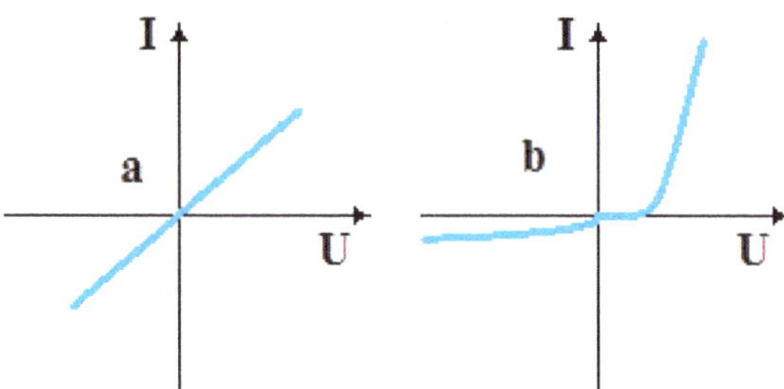

Fig. 6.21 Volt–Ampere characteristics of various types: (a) linear or Ohmic characteristic; (b) rectifying or diode characteristic.

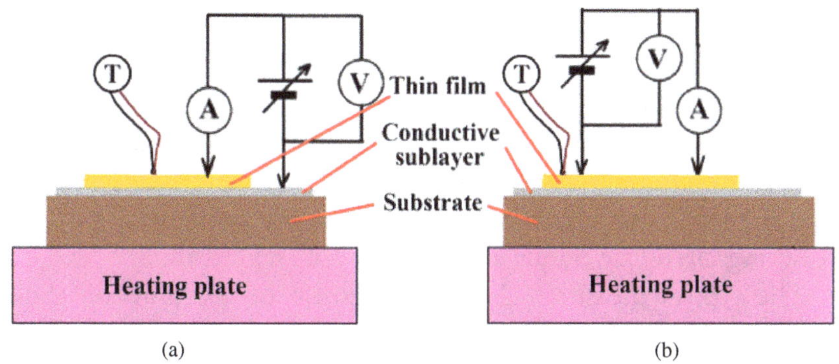

Fig. 6.22 Two shapes of the measuring scheme for the electrical characterization of thin films.

Figure 6.21 presents two types of *V–I* characteristics, a linear or Ohmic behavior (Fig. 6.21(a)) fully defined by Ohm's law and a rectifying or diode characteristic (Fig. 6.21(b)), which demonstrates the nonlinear behavior. When we evaluate the behavior of thin film under applied electric field, two possible measuring schemes may be realized. These schemes are shown in Fig. 6.22.

As shown in Fig. 6.22, in this way we can measure the transversal (a) and the lateral (b) electrical properties of thin films. Both schemes here are equipped with a heating plate and a thermometer, *T*, providing the possibility to take measurements at different temperatures. The main difference in these two ways is in the value of the applied electrical field affected on the thin film's material. For example, let us assume that the thickness of the film is 200 nm and the distance between electrodes in the case (b) is 1 cm. If we apply the supply of 5 V, the electrical field strength will be equal to $2.5 \cdot 10^5$ V/cm in the transversal case (Fig. 6.21(a)) and 5 V/cm in the case of the lateral electrical measurement. Evidently, the obtained results will be different due to the application of strong or weak electrical fields.

The evaluation method using measurements and building the *I–V* characteristics is a very powerful method. The shape of *I–V* characteristics is determined by the internal structure and properties of the measured thin film. However, a significant role belongs to the contact between the measuring electrode material and the studied thin film. Sometimes, the semiconductor pieces are used as the substrates for the studied film, which leads to the appearance of contact effects between materials. All these effects appear in the measured *I–V* characteristics. For example, Fig. 6.23 represents the *I–V* characteristics of two different devices: a junction diode (Fig. 6.23(a)) and a solar cell (Fig. 6.23(b)). Assume that both devices are connected with the measuring circuit by ohmic contacts.

The typical measured *I–V* characteristic of the junction diode consists of branches situated in two quadrants, first and third. The forward branch, in turn, consists of three regions: the cut-off region defined by the forward voltage $0 \le v \le V_\gamma$, the second region defined

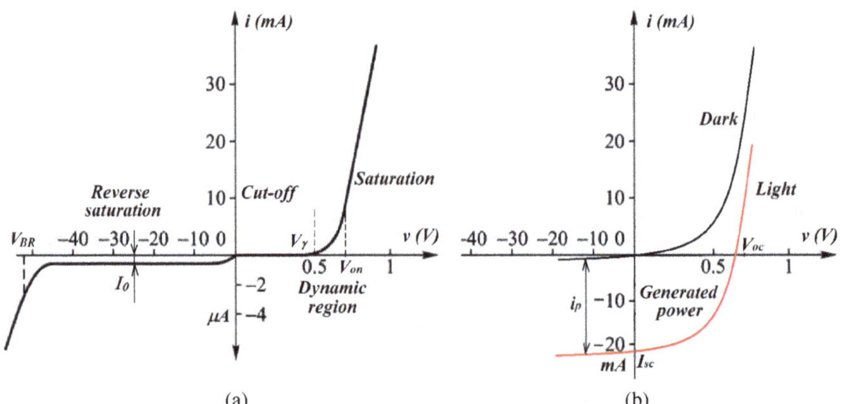

Fig. 6.23 Typical *I–V* characteristics of the junction diode (a) and a solar cell (b).

by $V\gamma \leq v \leq V_{on}$ called the dynamic region and the third region with $V_{on} \leq v$, which is called the saturation region. Here, v designates the applied voltage and $V\gamma$ and V_{on} are the approximate border voltages. In the cut-off region, the diode current is so small that it can be neglected due to the high internal resistance. The diode current begins to grow in the dynamic region due to decrease in the internal dynamic resistance of the diode. The following sharp increase of the diode current in the saturation region is practically independent of the applied voltage. In this region, the internal dynamic resistance attempts for a minimum. This behavior is described well by the well-known Shockley equation:

$$i = I_0\left(e^{\frac{V}{V_t}} - 1\right) \tag{6.49}$$

where $V_t = \frac{kT}{q}$ represents the ambient potential related with the absolute temperature of the environment in K and I_0 denotes the reverse saturation current of the diode. Internal dynamic resistance may be found using Eq. (6.49) as follows:

$$r = \frac{\partial V}{\partial i} = \frac{dV}{d\left[I_0\left(e^{\frac{V}{V_t}} - 1\right)\right]} = \frac{V_t}{I_0}e^{-\frac{V}{V_t}} \tag{6.50}$$

The reverse saturation current may be measured in the third quadrant of the I–V characteristics. The reverse branch situated in the third quadrant shows the reverse current as a function of the reverse voltage applied to the diode. A boundary value of the voltage leading to the irreversible breakdown is designated as V_{BR}. The reverse saturation current, according to the Shockley equation, depends on the three complex components. They are described by the following equation:

$$I_0 = qA\left(\frac{D_p p_{n0}}{L_p} + \frac{D_n n_{p0}}{L_n}\right) = qAn_i^2\left(\frac{D_p}{L_p N_D} + \frac{D_n}{L_n N_A}\right) \tag{6.51}$$

where A is the cross-section of the semiconductor junction, D_p and D_n are diffusivities of the holes and electrons respectively, L_p and L_n are diffusion lengths of the charge carriers. As is known, the diffusion length is defined by the lifetimes τ_p and τ_n of the charge carriers:

$$L_p \equiv \sqrt{D_p \tau_p} \text{ and } L_n \equiv \sqrt{D_n \tau_n} \tag{6.52}$$

Considering Eq. (6.51) together with Eq. (6.30) and taking into account the dependence of the diffusivity on temperature according to the Arrhenius law, one can write that the forward current of the junction diode also depends on temperature:

$$I_0 \propto T^{(3+\gamma)} e^{-\frac{E_g}{kT}} \rightarrow i_F \propto I_0 e^{\frac{v}{V_t}} \tag{6.53}$$

which is illustrated by Fig. 6.24, presenting the germanium diode I–V characteristics.

As shown in Fig. 6.24, the semiconductor junction diodes may be applied as temperature sensors.

Another type of I–V characteristic is presented in Fig. 6.23(b). The curve measured on the diode structure (the solar cell) under light

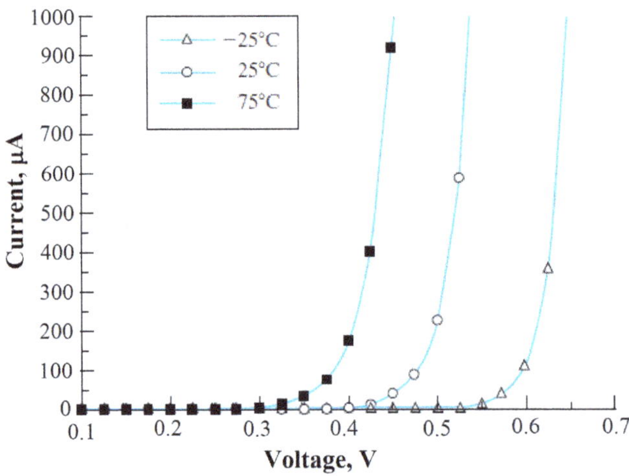

Fig. 6.24 The diode forward current dependence on the ambient temperature.

irradiation is in the fourth quadrant of the coordinate system. This shows that the product of coordinate measures, representing a power, will be negative. Therefore, the studied device produces energy unlike devices that have *I–V* characteristics situated in the first and third quadrants and absorbing the energy. In the presented *I–V* curve (see Fig. 6.23(b)), i_p denotes a current generated due to the light irradiation absorption, V_{oc} is the open circuit voltage generated by the solar cell and I_{sc} is the short-circuit current generated by the cell. The general equation describing the solar cell behavior differs from the Shockley equation on the value of the generated current.

$$i = I_0\left(e^{\frac{V}{V_t}} - 1\right) - i_p \tag{6.54}$$

The *I–V* characteristics showing the behavior of the photovoltaic element in the dark state and in the illuminated state are presented in Fig. 6.25.

General Eq. (6.54) enables us to calculate values of V_{oc} and I_{sc}. When we suppose that the general current i in Eq. (6.54) is equal to zero, we will obtain the formula for the V_{oc}:

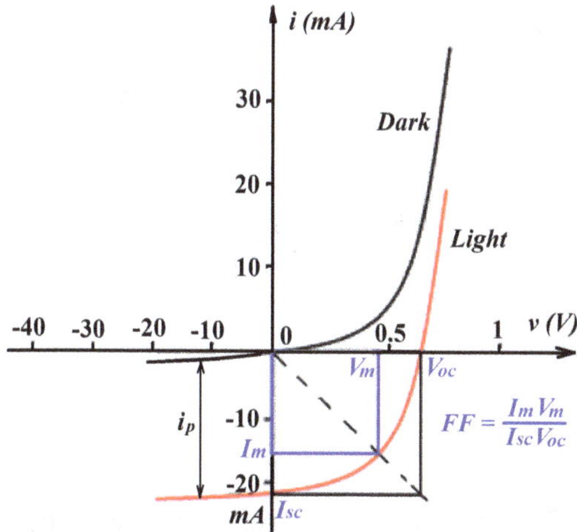

Fig. 6.25 Solar cell *I–V* characteristics measured in the dark and illuminated states.

$$V_{\mathrm{oc}} = V_t \ln\left(\frac{i_p}{I_0} + 1\right) \cong V_t \ln\frac{i_p}{I_0} \tag{6.55}$$

The value of the short-circuit current may be obtained from the same Eq. (6.54) making the generated voltage equal to zero and considering that $I_0 \ll i_p$:

$$I_{\mathrm{sc}} = -i_p \tag{6.56}$$

Product of these values, I_{sc} and V_{oc}, represents the power limit (the ideal power) that may be generated by a solar cell. This limit is shown in Fig. 6.25 as a square area $P_{\mathrm{id}} = V_{\mathrm{oc}} \cdot I_{\mathrm{sc}}$.

Effective generated power can be calculated as a product of generated current and voltage:

$$P = iV = I_0 V\left(e^{\frac{V}{V_t}} - 1\right) - i_p V \tag{6.57}$$

Differentiation of Eq. (6.57) and the condition $dP/dV = 0$ permit us to find the maximum generated voltage:

$$\frac{dP}{dV} = 0 = I_s\left(e^{\frac{V_m}{V_t}} - 1\right) - I_L + \frac{I_s V_m}{V_t} e^{\frac{V_m}{V_t}} \tag{6.58}$$

This transcendental equation can be solved for V_m using Eq. (6.55) as follows:

$$e^{\frac{V_m}{V_t}} = \frac{I_L + I_s}{I_s} \frac{1}{\frac{V_m}{V_t} + 1} \tag{6.59}$$

$$V_m = v_t \ln\frac{I_L + I_s}{I_s}\frac{1}{\frac{V_m}{V_t} + 1} = V_t \ln\frac{I_L + I_s}{I_s} - V_t \ln\left(\frac{V_m}{V_t} + 1\right) \cong V_{oc} - V_t \ln\left(\frac{V_m}{V_t} + 1\right) \tag{6.60}$$

If we substitute obtained value (6.60) into Eq. (6.54), we can find a maximum generated current I_m:

$$I_m \cong i_p\left(\frac{V_t}{V_m}-1\right) \tag{6.61}$$

Now, one can calculate a maximum generated power presented in Fig. 6.25 by multiplication of maximum generated voltage and current:

$$P_m = I_m V_m = i_p\left(\frac{V_t}{V_m}-1\right)\left[V_{\text{oc}} - V_t \ln\left(1+\frac{V_m}{V_t}\right)\right] = i_p\frac{E_m}{q} \tag{6.62}$$

where E_m is the energy that a photovoltaic cell can convert from the one absorbed photon to the electricity transferred to a load at the point of the maximum power. The relation between the maximum generated power and the ideal power is a significant parameter of the solar cell called a fill factor, *FF*:

$$FF = \frac{P_m}{P_{\text{id}}} = \frac{I_m V_m}{I_{\text{sc}} V_{\text{oc}}} \tag{6.63}$$

Using the fill factor, one can define an efficiency of the photovoltaic cell:

$$\eta = \frac{P_m}{P_{\text{in}}} = \frac{I_m V_m}{P_{\text{in}}} = \frac{FF \cdot I_{\text{sc}} V_{\text{oc}}}{P_{\text{in}}} = \frac{i_p E_m}{q P_{\text{in}}} \tag{6.64}$$

where P_{in} is the incident power of light. To measure the *I–V* characteristics of the type shown in Fig. 6.25, we use the measuring circuit presented in Fig. 6.26.

Figure 6.27 illustrates the practical realization of the measuring principal circuit shown in Fig. 6.26. This figure also presents a luxmeter applied for measurement of the light intensity and the electrode system for contacting with the metal electrode deposited on the semiconductor (silicon) surface.

Assembly and practical measurements are a part of the laboratory session devoted for evaluation of thin films and simple semiconductor devices prepared through the laboratory course.

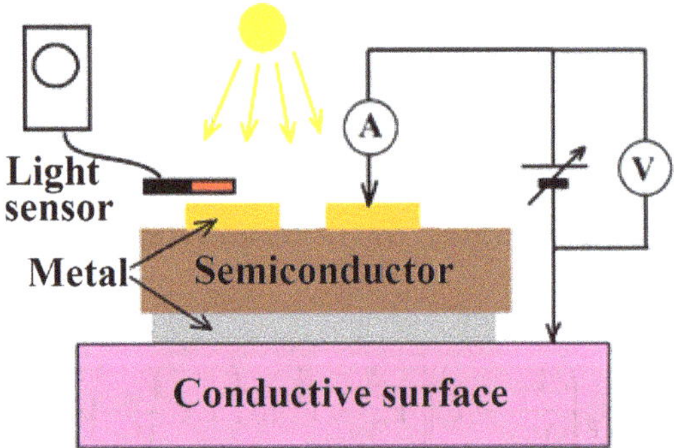

Fig. 6.26 The principal electric circuit for measurement of *I–V* characteristics under illumination.

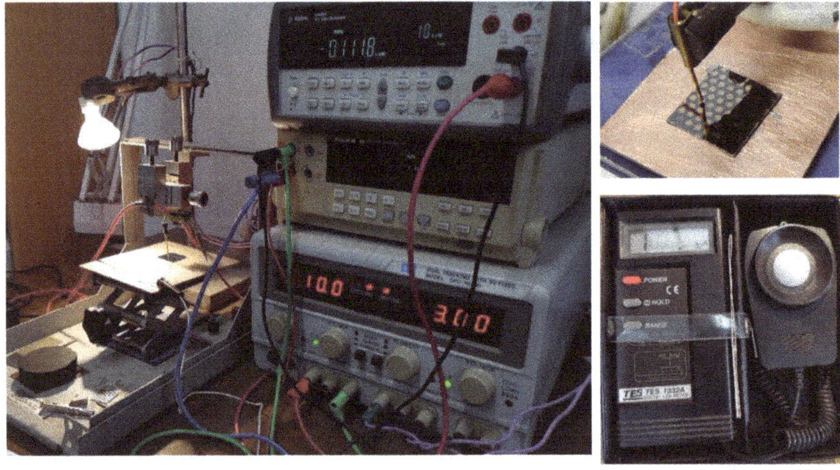

Fig. 6.27 Practical realization of *I–V* characteristics measurement under illumination.

A contact between a semiconductor and a metal thin film may be Ohmic and rectifying, as it was shown in Fig. 6.21. The rectifying type of a contact also appears at the connection of two different semiconductor materials, i.e., in the heterojunction case. The measured *V–I* characteristics enable evaluating the type of contact and the main mechanism of the charge carriers transport through this contact.

For example, let us consider the contact between metal and semiconductor. To liberate an electron from the metal, we need to apply

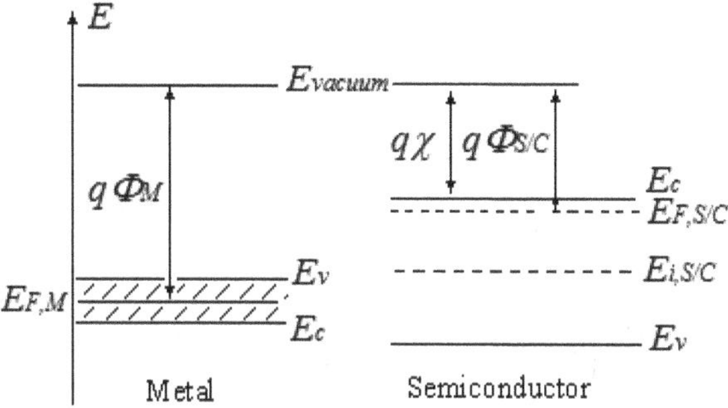

Fig. 6.28 Schematic illustration of metal and semiconductor in the energy coordinates.

to the metal an energy equal to the work function. On the other hand, to liberate an electron from the semiconductor, we also need to apply an energy equal to the work function, however the work function of semiconductors consists of two parts: the electron affinity or a width of the conduction band of the semiconductor and the distance between the Fermi level and the conduction band bottom, as shown in Fig. 6.28.

In this figure are shown two materials, metal and semiconductor. For metal, the work function is the energy value required for liberation of the electron from the metal's Fermi level outside, $q\Phi_M$. For semiconductor, the work function is a sum of the conductive band width ($q\chi$, electron affinity) and the distance between the Fermi level and the conductive band's bottom, $q\Phi_{S/C} = q\chi + (E_c - E_{F,S/C})$. Here, E_c designates the bottom of the conductive band, E_v designates the top of the valence band, $E_{i,S/C}$ is the Fermi level for an intrinsic semiconductor and $E_{F,S/C}$ is the Fermi level for the doped semiconductor. The semiconductor shown in Fig. 6.28 is of type *N* due to the proximity of the Fermi level to the conduction band and it has a work function lower than the work function of the metal.

When we bring these materials into contact, the process of transition to the thermodynamic equilibrium begins immediately. As a result, the Fermi levels of both materials should become equal. This leads to energy bands bending with the appearance of a potential barrier at the interface between the two materials. Figure 6.29 illustrates this process.

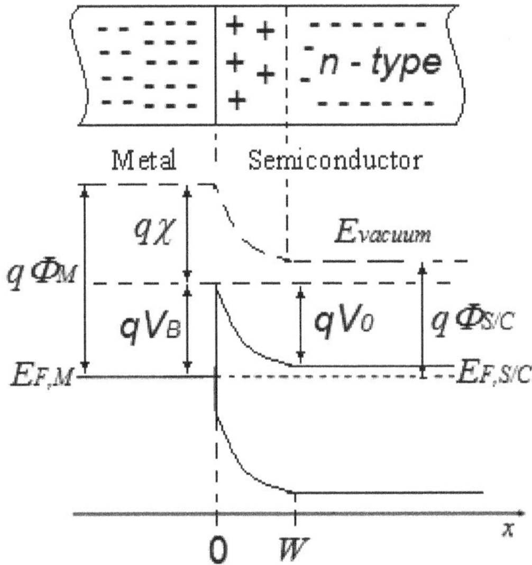

Fig. 6.29 Formation of a potential barrier at the interface between two media.

Now, free mobile electrons begin to cross the interface between materials in the direction of the metal due to the difference in the work functions. Thus, a depletion region forms on the semiconductor in the vicinity of the interface. This process continues until the appearance of a built-in electric field, which limits the movement of electrons from the semiconductor to the metal. This electric field appears due to the charge separation. By this way, the process of leveling the Fermi levels reaches a steady state. The width of the depletion zone, designated by value W, may be estimated in accordance with the theory of the junction diodes, considering that the difference in impurity concentration in metal and semiconductor is more than two orders. Thus, a value W may be estimated as follows:

$$W = \sqrt{\frac{2\varepsilon_0\varepsilon_r\left(V_0 - V_a\right)}{qN_D}} \tag{6.65}$$

where V_0 is the built-in potential, V_a is a voltage applied to the heterojunction structure and V_B is the potential barrier height.

To describe the electrical transfer through a rectifying contact, it is necessary to consider the mechanism of electron emission. There are several mechanisms for the transfer of an electron from one material to another. Each of them depends on different conditions. All mechanisms work together, but in each case one of them dominates. Basic mechanisms: diffusion current, thermionic emission (Schottky mechanism), tunneling (Fowler–Nordheim mechanism), emission from traps (Poole–Frenkel mechanism). The diffusion mechanism prevails in the case of Ohmic contacts and rectifying contacts, which are fed with reverse voltage. This reverse current is described as follows:

$$I_s = \frac{qAN_D D_n}{L_n} \tag{6.66}$$

where A represents the contact area, N_D is the impurity level in the semiconductor of N type, D_n and L_n are diffusivity and diffusion length of electrons. The tunneling mechanism arises due to the high concentration of impurities in the semiconductor, 10^{18}–10^{19} cm^{-3} and the wave properties of electrons. It is determined by the probability of electrons crossing the built-in potential barrier at the metal–semiconductor interface. If the current across the metal–semiconductor interface is proportional to the square of the ambient temperature under forward applied voltage, the current is called thermionic current. It flows, for example, in the case of an n-type semiconductor with a work function lower than that of a metal (see Fig. 6.29). This current obeys the following equation:

$$i = AA^*T^2 e^{-\frac{V_B}{V_t}} \left(e^{\frac{V_a}{nV_t}} - 1 \right) = I_0 \left(e^{\frac{V_a}{nV_t}} - 1 \right) \tag{6.67}$$

where n is the ideality coefficient and A^* is Richardson's coefficient, which is equal to

$$A^* = \frac{4\pi q m^* k^2}{h^3} = \left[\frac{A}{m^2 k^2} \right] \tag{6.68}$$

This contact is called a Schottky contact. In this case, the electron transport mechanism is defined by a thermo-electron emission of free charge carriers. A curve of measured *I–V* characteristics for metal–semiconductor contact regrown in the half-logarithmical form enables us to calculate the bandgap value and the ideality factor for the studied contact. Thus, the intersection of a derivative of the logarithm of the measured current with the voltage axis will bring the bandgap value, and an angle of this derivative defines the ideality factor:

$$V_B = V_T \ln\left(\frac{AA^*T^2}{I_0}\right); \qquad tg\alpha = \frac{d(\ln i)}{dV} = \frac{1}{nV_T} \tag{6.69}$$

Comparison of characteristics measured for different metals shows what metal displays the Schottky contact with the used silicon substrate. In the presence of an external electrical field, the potential barrier between metal and semiconductor decreases and Eq. (6.67) transforms in the following form:

$$i = AA^*T^2 e^{-\frac{\left(qV_B - \sqrt{\frac{qE}{\pi\varepsilon_0\varepsilon_r}}\right)}{V_t}} \left(e^{\frac{V_a}{nV_t}} - 1\right) \tag{6.70}$$

The effect of decreasing the potential barrier height under influence of the applied electrical field is known as a lowering effect. When the applied electrical field increases, an additional mechanism of the electron emission appears. This mechanism, known as a Poole–Frenkel mechanism, represents the emission from the structural defects that are energetic traps distributed in the semiconductor body or in the interface between metal and semiconductor. This mechanism is described as follows:

$$i = An\mu qEe^{-\frac{\left(qV_B - \sqrt{\frac{qE}{\pi\varepsilon_0\varepsilon_r}}\right)}{V_t}} \left(e^{\frac{V_a}{nV_t}} - 1\right) \tag{6.71}$$

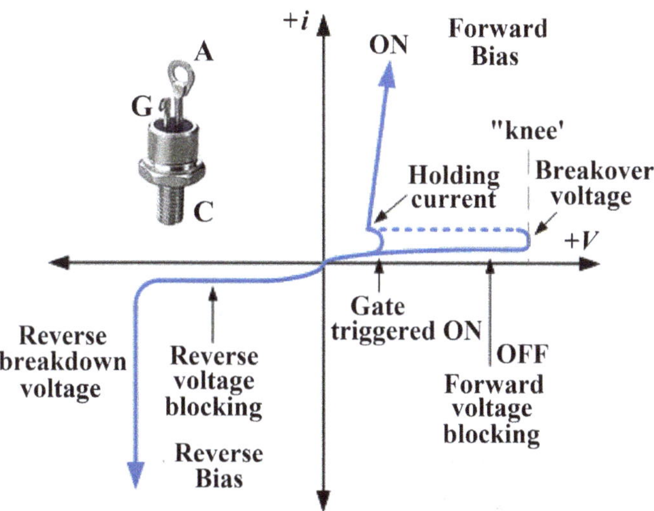

Fig. 6.30 *I–V* characteristic of a controlled diode (thyristor).

where n is the carriers' concentration and μ is the mobility of the carriers. Evidently, we can observe the difference between both mechanisms, Schottky and Poole–Frenkel, when comparing the *I–V* characteristics behaviors.

Figure 6.30 presents one more type of *I–V* characteristics, showing a nonlinear curve which has a "knee" conditioned by the device's (thyristor) behavior.

A thyristor represents a four-layer semiconductor device. Its equivalent circuit may be presented as two bipolar transistors of p- and n-types connected in Darlington's scheme. This circuit enables obtaining very high gain, which provides an avalanche type of work with very quick switching in the conduction state. Their equivalent circuit and a schematic view of the thyristor are shown in Fig. 6.31.

A thyristor has three external electrodes like the transistor. However, the working principle of these devices is very different. In the transistor, we can control the value of a current flowing through the transistor by changing the control voltage on its base. In the thyristor, we can control the time when the thyristor conducts a current by application of the control constant voltage or a short voltage pulse

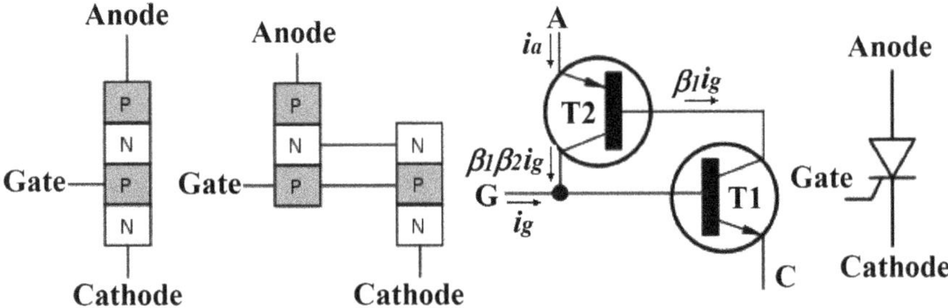

Fig. 6.31 Equivalent circuit of the thyristor.

to the control electrode, the gate. Moreover, a thyristor may be turned on only if an anode voltage is positive with respect to the cathode. This behavior is illustrated by the *I–V* curve shown in Fig. 6.30. Here, a control voltage pulse may be applied only at the positive voltage bias between anode and cathode. Thus, the "knee" shown on the curve, designates the turn-on switching and sharp growth of the forward current which is reached after application of the control pulse. This is possible only in the positive part of the curve in the first quadrant. At negative voltage on the anode relative to the cathode, the thyristor is in the turn-off state.

6.4 Capacitance and Volt–Farad characterization

One of the important parameters of the metal–semiconductor junction is the capacitance of the junction. The system made of a metal–oxide–semiconductor–metal structure represents a MOS capacitor. In addition, in the structure of a conventional junction diode, there is a corresponding capacitance formed due to the formation of a depletion region at the semiconductor junction. This capacitance determines the speed of semiconductor devices and allows us to understand the processes taking place in these devices. The depletion region width in the case of a Schottky diode is defined by Eq. (6.65). The depletion region in the P–N junction diode is formed by diffusion currents that occur in the metallurgical border between two semiconductors with different charge carriers concentrations. Figure 6.32 illustrates the depletion regions' formation in the junction diode.

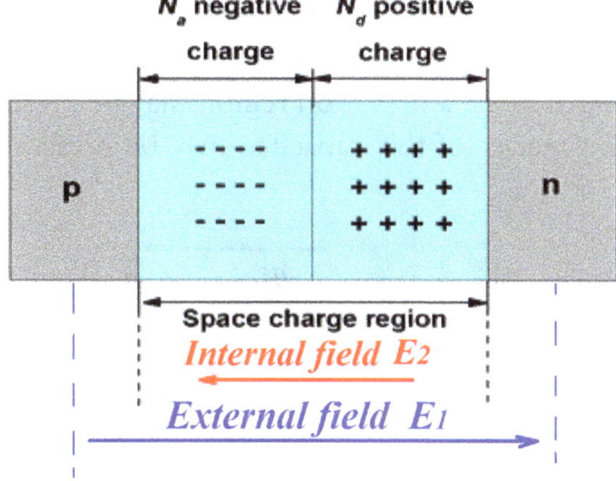

Fig. 6.32 Formation of the depletion region in the junction diode.

When we close together two semiconductors of P- and N-types, a natural diffusion process begins due to the external electrical field created by two parts of the semiconductor. Positive charge carriers diffuse from the P-type part to the N-type part and negative charge carriers move in the opposite direction forming in this way an internal electric field. The steady state will be reached when $E_1 = E_2$. Thus, the depletion regions will be formed in both contacting parts of the semiconductor. The built-in potential in the common depletion region is equal to

$$V_0 = V_t \ln \frac{N_D N_A}{n_i^2} \tag{6.72}$$

where N_A and N_D are the acceptor and donor concentrations, respectively, and n_i is the intrinsic concentration in the applied semiconductor. The width of the depletion region is described like the same in the Schottky diode, considering both parts of it:

$$W = \sqrt{\frac{2\varepsilon_0 \varepsilon_r (N_D + N_A)(V_0 - V_a)}{q N_D N_A}} \tag{6.73}$$

As can be seen from Eqs. (6.65) and (6.73), the width of the depletion region depends on the value of an external applied voltage V_a.

A depletion region is characterized by absence of mobile charge carriers, therefore its resistance is very high, and in the case of a sharp doping from both sides, a depletion region may be presented as a flat capacitor. The capacity of this capacitor may be presented as follows (see Eq. (6.17)):

$$C_j = \frac{\varepsilon_s A}{W} = \frac{\varepsilon_0 \varepsilon_r A}{W} = A \cdot \sqrt{\frac{q\varepsilon_0\varepsilon_r}{2(V_0 - V_a)} \frac{N_A N_D}{N_A + N_D}} \tag{6.74}$$

In the particular case, $N_A >> N_D$, this equation is simplified as follows:

$$C_j \approx A \cdot \sqrt{\frac{q\varepsilon_s N_D}{2(V_0 - V_a)}}; \frac{1}{C_j^2} = \frac{2(V_0 - V_a)}{qN_D \varepsilon_s A^2}; \frac{1}{C_j^2} = \frac{2}{qN_D \varepsilon_s A^2} V_0 - \frac{2}{qN_D \varepsilon_s A^2} V_a \tag{6.75}$$

This equation describes a plot of the direct line. An intersect of this line with the abscissa axis presents the bandgap V_0 of the studied semiconductor and the slope of the direct line defines the major carriers concentration value:

$$N_D = -\frac{2}{q\varepsilon_s A^2 tg\alpha} \tag{6.76}$$

This analysis is illustrated in Fig. 6.33.

A metal–semiconductor junction or a metal–insulator–semiconductor junction behave similar to the devices shown in Fig. 6.32. Such structure is presented schematically in Fig. 6.34. Here, to be specific, the metal–oxide–silicon structure is presented. The applied metal (the upper electrode) forms the Schottky junction with the silicon of P-type. The lower electrode forms the Ohmic contact with the silicon.

Without the applied voltage V_a, the thin film system goes to the steady state with the formation of the depletion region defined by the impurities concentration in the semiconductor. In the thermodynamical steady state, the Fermi levels in the metal and semiconductor align, which leads to curvature of the energy bands as shown in Fig. 6.35.

As shown in these diagrams, built for the case of a p-type semiconductor and a metal with work function lower than the same of the

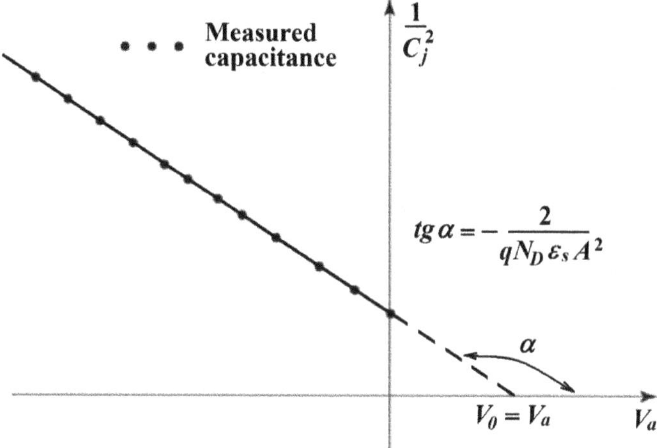

Fig. 6.33 A junction capacitance behavior under applied external voltage.

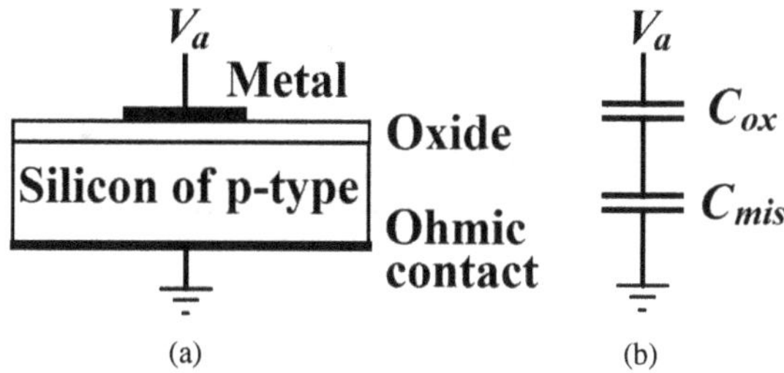

Fig. 6.34 The schematic view of the metal–oxide–silicon structure (a) and the equivalent circuit of this structure represented by two capacitors connected in series.

semiconductor, the borders of conduction and valence bands bend and build the potential barrier for the charge carriers transport between contacting materials. The thickness of the depletion region depends on the external applied voltage V_a (see Eq. (6.73)). Therefore, the systems metal–semiconductor and metal–insulator–semiconductor (MIS) may be considered as variable capacitors. Properties of these capacitors may be found from Volt–Farad characteristics which should be plotted using measurement results. Further, such parameters of these Schottky diodes as the built-in potential and the major charge carriers concentration will be calculated from measured data of the junction capacitance.

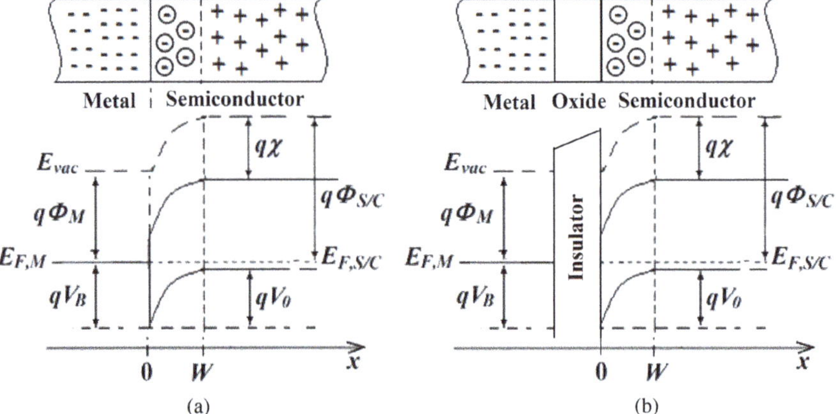

Fig. 6.35 The schematic view of the energy band diagrams for two Schottky contacts: (a) the metal–semiconductor structure and (b) the metal–insulator–semiconductor structure.

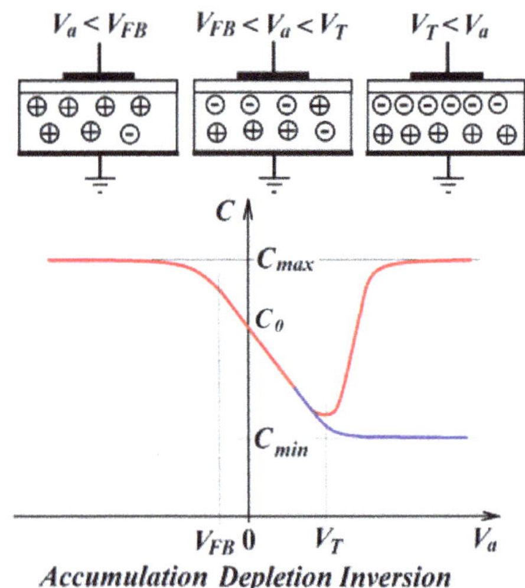

Fig. 6.36 Ideal Volt–Farad curve plotted for the MIS capacitor with a p-type semiconductor.

Let us consider dipper, what happens when we apply the external voltage to the MIS structure. For simplicity, suppose that the area of our structure is equal to 1 cm^2. Figure 6.36 illustrates the behavior of the MIS structure under variable V_a.

The studied structure represents two connected in series capacitors C_{ox} and $C_{mis} = C_j$ as shown in Fig. 6.34(b), where C_{ox} is the specific capacity of the constant capacitor based on the insulator or oxide layer situated between the metal contact and the semiconductor surface and C_j is the specific capacity of the variable capacitor defined by the depletion region width. Therefore, the total capacity C_Σ will be equal to

$$\frac{1}{C_\Sigma} = \frac{1}{C_{ox}} + \frac{1}{C_j} \rightarrow C_\Sigma = \frac{C_{ox}C_j}{C_{ox} + C_j} = \frac{C_{ox}}{\frac{C_{ox}}{C_j} + 1} \tag{6.77}$$

The specific capacity of the dielectric layer is equal to

$$C_{ox} = \frac{\varepsilon_0 \varepsilon_i}{d_{ox}} \tag{6.78}$$

where ε_i is the permittivity of the insulator layer and d_0 is its thickness. The specific capacity of the MIS junction is determined by the following formula (see Eq. (6.75)):

$$C_j = \frac{\varepsilon_0 \varepsilon_r}{W} \approx \sqrt{\frac{q\varepsilon_s N_A}{2(V_0 - V_a)}} \tag{6.79}$$

Substituting Eqs. (6.77), (6.78) and (6.79), we will obtain the total specific capacity as follows:

$$C_\Sigma = \frac{\varepsilon_0 \varepsilon_i}{d_{ox}\left(\frac{\varepsilon_0 \varepsilon_i W}{d_0 \varepsilon_0 \varepsilon_r} + 1\right)} = \frac{\varepsilon_0 \varepsilon_i}{W\frac{\varepsilon_i}{\varepsilon_r} + d_{ox}} = \frac{\varepsilon_0 \varepsilon_i}{d_{ox} + \varepsilon_i \sqrt{\frac{2\varepsilon_0 (V_0 - V_a)}{\varepsilon_r q N_A}}} \tag{6.80}$$

Analysis of this equation together with Fig. 6.36 explains the behavior of the system. When the applied voltage V_a is negative ($V_a = -V_R$) and lower than the flat-band voltage V_{FB}, the positive charge carriers in the semiconductor are attracted to the semiconductor–insulator interface. This process is called the accumulation process.

The system turns into a capacitor with capacity defined by the insulator layer only:

$$C_\Sigma \approx C_{ox} = C_{max} = \frac{\varepsilon_0 \varepsilon_i}{d_{ox}} \tag{6.81}$$

When the voltage V_a increases in the range $V_{FB} < V_a < V_T$, positive mobile charge carriers are squeezed out of the semiconductor–insulator interface, forming by this way a depletion region. The total capacitance of both capacitors connected in series is now described by Eq. (6.80). Here, the voltage V_T is the threshold voltage that ensures the concentration of negative charge carriers near the semiconductor–insulator interface equal to the concentration of acceptor impurities, $n = p$. The state with applied voltage exceeding the threshold voltage is called an inversion state. A negative space charge now exists at the boundary of the insulator, and the capacity of the system returns again to the same Eq. (6.81), which is illustrated by the red line in Fig. 6.36. In the case of measurement at high frequency, the built space charge cannot react synchronously with the applied signal, the depletion region keeps its boundaries and the capacity of the system reaches the minimum value, C_{min}, as shown in Fig. 6.36 by the blue line. Figure 6.37 presents the schematic principle measuring circuit for C–V characterization. This circuit consists of a function generator, a bias supply, an oscilloscope and a measuring setup. The main requirements to the measuring setup are the fixed resistance between the measuring probe and the oscilloscope input, usually 50 Ω. Connection of the measuring probes with the oscilloscope most be prepared with the coaxial cables. A bias voltage varies in the

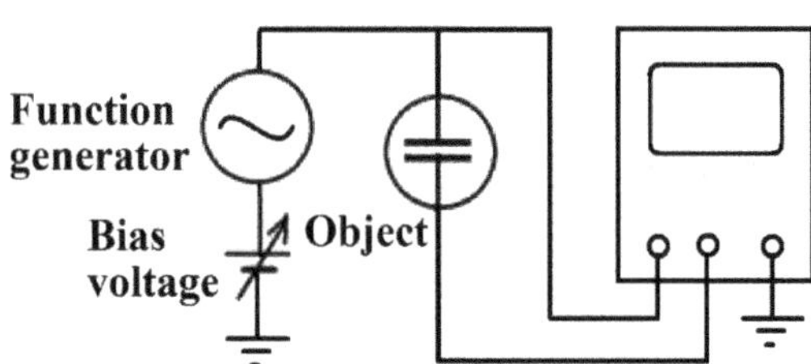

Fig. 6.37 A principle measuring scheme for C–V characterization.

interval of –5 V up to +1 V. The measuring signal represents a sinusoidal or a square wave with frequency of 0.1–1 MHz and amplitude of 20–100 mV.

Let us consider Eq. (6.74) once more. The junction capacitance depends on the applied voltage. When the applied voltage is negative, the reverse voltage V_R, the junction capacitance decreases and reaches its minimum value, while the width of the depletion region reaches the size of the diode. Under the positive voltage, the junction capacitance increases and disappears when the applied voltage reaches the built-in potential value. In this case, the depletion region also disappears and current continues to flow through the diode.

According to the general equation defining the electrical charge as a number of charge carriers flowing during a defined time through the semiconductor junction and charging it, one can write:

$$Q = it = CV \tag{6.82}$$

Thus, if the current flows through the diode under a forward voltage, one can define the diffusion capacitance of the diode as follows:

$$C_D = \frac{d\Delta Q}{dV_a} = \tau_{p,n} \frac{di}{dV_a} \tag{6.83}$$

where C_D designates the diffusion capacitance, ΔQ is the charge passed through the diode and $\tau_{p,n}$ is the time required for passing through the diode for the charge carriers. As a first approximation, a time $\tau_{p,n}$ may be defined using a diffusion length and a diffusivity that is suitable for the charge carriers:

$$\tau_{p,n} = \frac{L^2}{D} \tag{6.84}$$

Now, substituting the Shockley equation (Eq. (6.49)) into the Eq. (6.83), we can find the value of the diffusion capacitance and their relation with the applied voltage.

$$C_D = \tau_{p,n} \frac{di}{dV_a} = \tau_{p,n} \frac{dI_0\left(e^{\frac{V_a}{V_t}} - 1\right)}{dV_a} = \frac{\tau_{p,n} I_0}{V_t} e^{\frac{V_a}{V_t}} \quad (6.85)$$

We will consider the physical meaning of the diffusion capacitance using Fig. 6.38.

Diffusion capacitance in a diode exists only if there is a forward voltage at the junction and the current flows through it. One can imagine a diffusion capacitor as a virtual flat capacitor when the lower edge of the conduction band and the upper edge of the valence band play a role of the flat capacitor electrodes and the forbidden gap keeps the same distance between these electrodes in the P-type part as in the N-type part of the diode. When a forward voltage is applied to the junction, electrons move to the positive electrode through the conduction band and holes move to the negative electrode through the valence band, forming the diffusion capacitance. Under the reverse voltage on the junction, the diffusion capacitance decreases such that it may be neglected.

Figure 6.39 presents the simplest way to measure the capacity of the metal–semiconductor junction using a function generator, a decade variable resistor, an oscilloscope and a measuring setup.

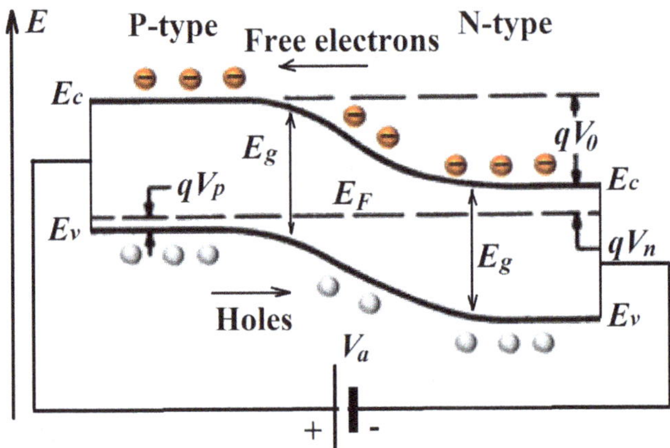

Fig. 6.38 Principal scheme of the diffusion capacity in the energy coordinates.

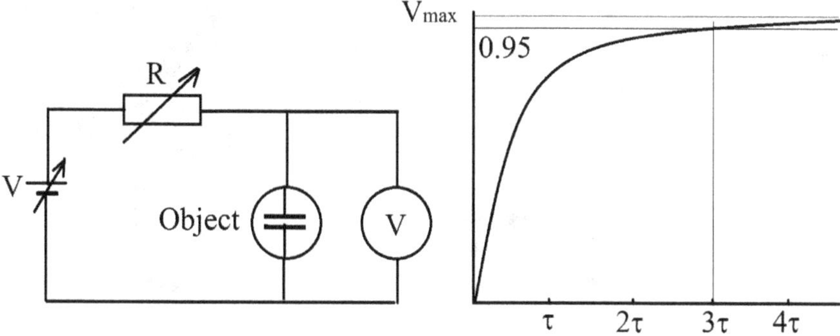

Fig. 6.39 The simplest circuit for measuring the capacitance of a semiconductor junction.

A function generator here represents the supply. To provide the measuring process, the square wave from the function generator with 5 V amplitude, 2.5 V positive bias and frequency 1 kHz should be applied to the measured sample through the limiting resistor of R = 1 kΩ. Two-beam oscilloscope was used as the voltmeter, the first channel is conducted with the generator output and the second channel presents the studied voltage on the sample.

The capacitance may be approximately calculated using the time required for charging the measured semiconductor structure. It is known that the capacitor charges up to 95% of its maximum charge through a time equal approximately to 3τ, where τ is the time constant related with capacity as follows: $\tau = RC$. Now, one can calculate the capacitance using the obtained oscilloscope's plot and the following equation:

$$C = \frac{\tau}{R} \tag{6.86}$$

Figure 6.40 illustrates this type of measurement.

As shown in Fig. 6.40(a), the measuring system consists of the function generator, the decade variable resistors, the oscilloscope and the sample's arrangement. The sample arrangement includes a conductive plate, the studied sample and the spring-loaded electrode as presented in Fig. 6.40(b). The picture shown in Fig. 6.40(c) enables

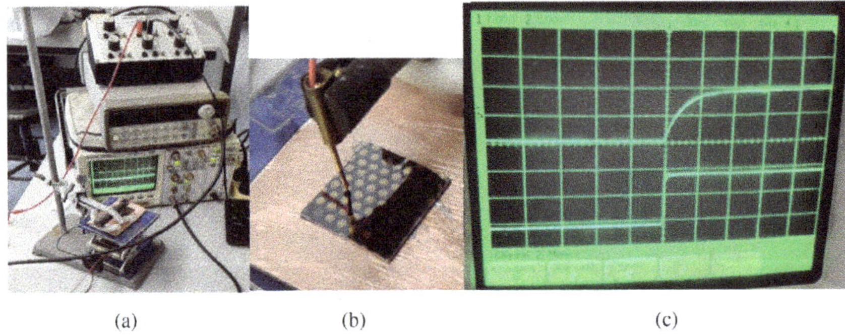

(a) (b) (c)

Fig. 6.40 Practical realization of a capacitance evaluation: the measuring devices (a), the sample measuring arrangement (b) and the photography of the measured curves (c).

us to measure the experimental value of the time constant and calculate the capacity. It should be noted that in this way we measure and calculate the diffusion capacitance, C_D, of the metal–semiconductor structure. As such presented measurements are conducted at the positive applied voltage only, this voltage significantly exceeds the built-in potential of the junction.

To measure the junction capacitance, C_j, and its dependence on the voltage bias, we need to apply the more complex measuring circuit. According to Eqs. (6.74) and (6.75), this capacitance depends on the applied voltage V_a and exists in the range $V_{BR} < V_a < V_0$, where V_{BR} represents the reverse break-down voltage and V_0 represents the built-in potential. The applied measuring method is called the Volt–Farad characteristics method. A distinctive feature of this method is using of two electrical signals applied to the measured structure. First one is the constant bias voltage, providing the required thickness of the depletion region in the interface metal–semiconductor. The second is the alternating voltage of the small amplitude, the measuring signal.

6.5 Dynamical Hot-Probe characterization

The main parameters of the basic crystalline semiconductors are well known and are found in many reference books. These parameters are the type of semiconductor, impurity concentration, carrier mobility and diffusion coefficient. Hall Effect measurements provide data on the mobility and concentration of majority of the carriers. The Haynes–Shockley experiment provides data on the minority carrier mobility and diffusion coefficient. The main problems in application of these methods to thin films parameters measurement are in the measuring devices' sensitivity and in the complexity of specific samples' preparation. Samples for Hall Effect measurement require high-precision electrodes arrangement. The Haynes–Shockley experiment necessitates specific equipment with short switching time: an electrical pulse generator or a pulse laser for minority carrier's excitation. In addition, the equipment for Hall and Haynes–Shockley measurements is very expensive, so these methods cannot be applied for fast evaluation of thin films. The main requirements for assessment methods are simplicity, efficiency and sufficient accuracy. The problem may be solved using the modified hot-probe experiment.

The traditional hot-probe experiment provides a simple and effective way to distinguish between n-type and p-type semiconductors using a hot probe and a standard multimeter. A principal scheme of the experiment is presented in Fig. 6.41.

Here, a pair of cold and hot probes are attached to the semiconductor surface. The hot probe connects to the positive terminal of the multi-meter and the cold probe to the negative terminal. Applying

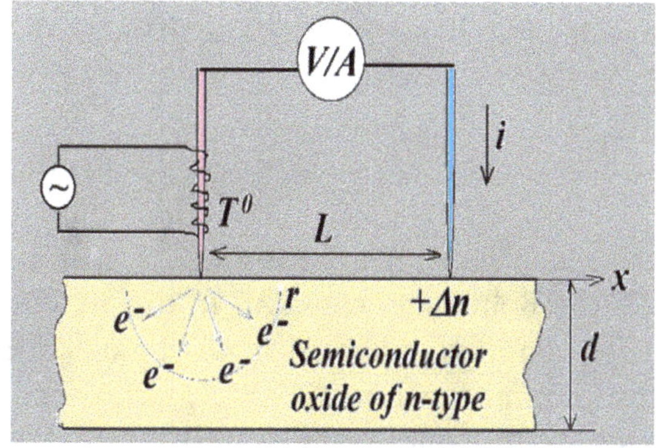

Fig. 6.41 A principal measurement scheme of the hot-probe experiment.

cold and hot probes to an n-type semiconductor produces a positive voltage on the meter, while a p-type semiconductor produces a negative voltage. The explanation of this experiment is that thermally excited free majority charge carriers move within the semiconductor from a hot probe to a cold one. The mechanism of this movement inside the semiconductor is diffusion, since heated charge carriers move faster than cold ones. These transferred majority carriers determine the sign of the electrical potential of the measured current in the multi-meter. If we short-circuit the hot and the cold electrodes with an ammeter, a current will begin to flow through it. The sign of the current shows the type of the major charge carriers also.

The modified or dynamical hot-probe measurement technique distinguishes from the usual by the time dependent recording of voltage or current measured during the hot-probe experiment. The experiment takes only 30–90 s in order to exclude the effect of heating the second (cold) electrode. The diagrams recorded during the hot-probe experiment enable us to calculate the basic parameters of the semiconductor thin films. Evidently, the calculations take into account the measured values of such parameters as the sheet resistance measured by the four-point method and the thickness of the studied thin film as well.

Charge carriers in semiconductors can be in two states: in a state of static (thermodynamic) equilibrium and a non-equilibrium state due to external influences, which can be electric, magnetic or temperature fields. The distribution of charge carriers in energy at equilibrium is determined by the Fermi–Dirac equation

$$f_0(E) = \frac{1}{1 + e^{\frac{E - E_F}{k_B T}}} \tag{6.87}$$

where E_F is the Fermi energy, T is the absolute temperature in degrees Kelvin and k_B is the Boltzmann constant. In the presence of external influences, the system of charge carriers will be described by a non-equilibrium distribution function $f(E) = f(p,r,t)$, which depends on energy (momentum, p), coordinates (r) and time (t). This function, $f(E)$, includes all possible mechanisms by which the distribution

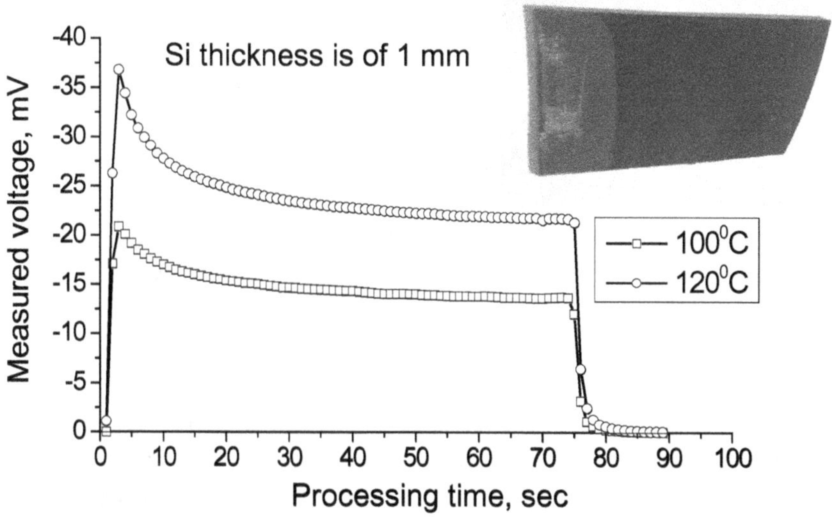

Fig. 6.42 Dynamical hot-probe characteristics recorded for the bulk silicon sample of p-type at different temperatures; the inset represents the measured sample.

function can be changed. In real materials, all charge carriers are subject to collisions with other particles, impurities, phonons, etc. Thus, the function $f(E)$ must take into account the scattering of charged carriers due to collisions when the function passes into another equilibrium state. If the external influence stops, the function should return to its original state. Therefore, the behavior of the distribution function can be represented by three different parts: excitation, determined by the magnitude of the influencing external field; steady state, with slow ongoing processes; and a relaxation part, with a return to its original state. Figure 6.42 illustrates these three parts of the recorded dynamical hot-probe diagram.

Figure 6.42 represents recorded hot-probe characteristics measured on the bulk silicon sample of p-type. The distance between electrodes was 1 cm. A heater was contacted with one of the fixed electrodes placed on the sample surface. After several tens of seconds, the heater was removed from the sample.

All these parts may be described by the same general equation, taking into account all external and internal processes. In general, the behavior of charge carriers, in particular electrons, can be described by the well-known BTE, which in differential form looks like this (see Eq. (6.14)):

$$\frac{\partial f}{\partial t} + \dot{k}\frac{\partial f}{\partial k} + \dot{r}\frac{\partial f}{\partial r} = \left(\frac{\partial f}{\partial t}\right)_{\text{Coll}} \tag{6.88}$$

where first summand on the left-hand side represents the function variation in time, the second represents influences of external fields and third shows the coordinate variation. The right part of the equation represents various types of collisions influencing the particles' movement: scattering on ionized impurities, scattering on neutral impurities, scattering on dislocations and scattering on grain barriers that are very important for polycrystalline semiconductor thin films. The right-hand side represents a scattering of the particles and, for simplicity, it may be presented using an electron relaxation time, τ. In the case when there are no external fields and charge carriers, return from the disturbed state to the initial state occurs as follows (relaxation time approximation):

$$\left(\frac{\partial f}{\partial t}\right)_{\text{Coll}} = -\frac{f - f_0}{\tau} \tag{6.89}$$

This value, τ, depends on the dominant scattering mechanisms and may be found from experimentally measured dynamical hot-probe characteristics, see, for example, the almost exponentially decreasing voltage on Fig. 6.42. This decreasing voltage illustrates the relaxation process.

For simplicity, we convert the BTE (Eq. (6.88)) to an equation without external electrical and magnetic fields. Now only the temperature gradient affects the movement of charge carriers. The internal electric field arises due to a directional flow of charge carriers. We also assume that the material being measured is isotropic and the charge carriers flux occurs in only one x direction, where x denotes the r coordinate. Momentum of a free electron may be described using the relation $p = mv = \hbar k$, where $\hbar$ is the reduced Planck constant, k is the wave vector, m is the mass of electron and v is the velocity of the electron. Electrons can move due to the Coulomb electrical force F:

$$F = m\frac{dv}{dt} = \hbar\frac{dk}{dt} = -qE_{\text{ex}} \tag{6.90}$$

where E_{ex} is an external electrical field. According to the relation (6.90), the force related term in the full BTE (Eq. (6.88)) may be described as follows:

$$\dot{k}\frac{\partial f}{\partial k} = -qE_{\text{ex}}v\frac{df}{dE} \tag{6.91}$$

However, in the case of no external electrical field, this term will vanish.

When the material is heated at the certain point on the surface, the charge carriers begin to move and, thus, they create a current and an electrical field in accordance with the following relationship:

$$E_x = \frac{1}{\sigma}j + \beta\frac{dT}{dx} \tag{6.92}$$

where j is the current density, σ is the conductivity of a material depended on temperature and β is the additional coefficient characterizing the thermo-electrical properties of the material. This coefficient represents thermopower (the Seebeck coefficient) produced in the material under non-homogeneous heating due to the charge carrier's transport. At the same time, the heated electrode excites additional charged carriers which are the free charged carriers. This process is different for various materials and various temperatures and it significantly depends on the bandgap of the semiconductor. We can estimate an influence of the thermally excited additional charge carriers on the system behavior.

Let us consider once more Eq. (6.30), which describes the charge carriers concentration as a function of temperature:

$$n_i = AT^{\frac{3}{2}}e^{-\frac{E_g}{2kT}} \tag{6.93}$$

If we will designate the normal temperature (RT) as T_0, the number of charge carriers at elevated temperature will be n_{iT}:

$$n_{iT} = n_i \left(\frac{T}{T_0}\right)^{\frac{3}{2}} \exp\left[\frac{E_g}{2kT_0}\left(1 - \frac{T_0}{T}\right)\right] = n_i R, \text{ and}$$

$$R = \left(\frac{T}{T_0}\right)^{\frac{3}{2}} \exp\left[\frac{E_g}{2kT_0}\left(1 - \frac{T_0}{T}\right)\right] \tag{6.94}$$

where R is a function of the heating temperature. Therefore, the additional excited charge carriers for the p-type semiconductor will be defined by the following equation (see Eqs. (6.35) and (6.36)):

$$\Delta n = \Delta p = \sqrt{\left(\frac{N_A}{2}\right)^2 + n_i^2 R^2} - \sqrt{\left(\frac{N_A}{2}\right)^2 + n_i^2} \tag{6.95}$$

This equation may be approximately solved for Δn. As such $\Delta n < n_i < n_i R$ in our assumptions, one can neglect small values and the solution we obtain is in the following form:

$$\Delta n \approx \frac{n_i^2 R^2}{N_A} = \frac{n_i^2}{N_A}\left(\frac{T}{T_0}\right)^3 \exp\left[\frac{E_g}{kT_0}\left(1 - \frac{T_0}{T}\right)\right] \tag{6.96}$$

This equation makes it possible to estimate the growth of the intrinsic carrier concentration with temperature depending on the semiconductor bandgap. However, if we compare usual semiconductors such as silicon and wide-bandgap semiconductors with bandgap of 2–4 eV, one can see that their intrinsic concentration at room temperature is insignificantly small, see, for example, Fig. 6.43. Therefore, only thermal running away of the heated charge carriers may be taken into account through the evaluation process of wide-bandgap semiconductors at temperatures lower than 250°C.

Let us return to Eq. (6.92), in the case that the current reaches the steady state under heating:

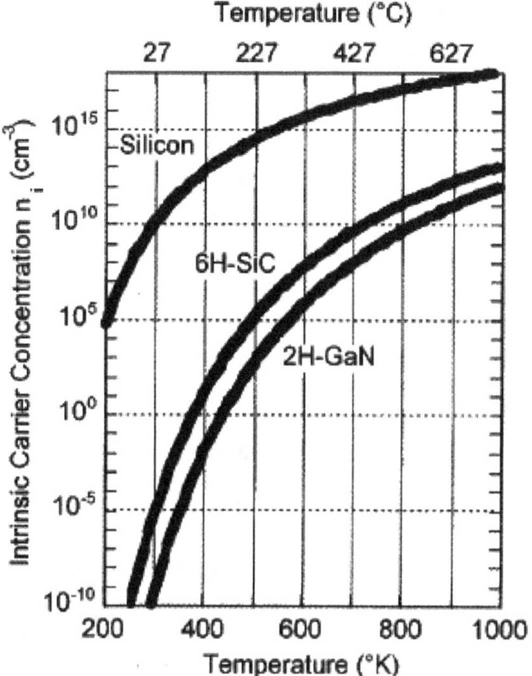

Fig. 6.43 Intrinsic charge carrier concentration dependencies on temperature for semiconductors with narrow and wide-bandgap materials.

$$E_x = \frac{1}{\sigma} j_s + \beta \frac{dT}{dx} \tag{6.97}$$

where j_s represents the saturation current. This equation can be integrated on the distance L (see Fig. 6.40), that brings the following result:

$$V_L = \frac{L}{\sigma} j_s + \beta \Delta T \tag{6.98}$$

where V_L is the voltage drop that appears as a result of charge carriers separation. The conductivity of the sample may be described by the known formula (the Drude equation, see Eq. (6.41)):

$$\sigma = \frac{1}{\rho} = q\mu n \tag{6.99}$$

Both parameters, mobility and concentration of charge carriers, are functions of temperature. Mobility of charge carriers, μ, may be found using the well-known Einstein relation:

$$D = \mu \frac{kT_e}{q} \tag{6.100}$$

where D is the diffusion coefficient of the major charge carriers, k is Boltzmann's constant and T_e is an environment temperature in K. Taking into account that the diffusion distance is equal to $L = \sqrt{D\tau}$, where τ is the measured relaxation time and substituting with Eq. (6.100), we obtain

$$\mu = \frac{L^2 q}{k_B T_e \tau} \tag{6.101}$$

Relaxation time may be found from recorded hot-probe characteristics using a numerical differentiation method, for example, using the known approximate three-point formula of Lagrange:

$$\begin{cases} f'(t_0) = \frac{1}{2\Delta t}\left[-3f(t_0) + 4f(t_0 + \Delta t) - f(t_0 + 2\Delta t)\right] \\ f'(t_0 + \Delta t) = \frac{1}{2\Delta t}\left[-f(t_0) + f(t_0 + 2\Delta t)\right] \\ f'(t_0 + 2\Delta t) = \frac{1}{2\Delta t}\left[f(t_0) - 4f(t_0 + \Delta t) + 3f(t_0 + 2\Delta t)\right] \end{cases} \tag{6.102}$$

where a function f represents the measured voltage, t_0 is the first time-point of the decreasing function and Δt is the time–space between measured points (see the decreasing part of diagram in Fig. 6.42, for example). The thermopower may be found by the same way for suitable temperatures.

Combining Eqs. (6.98) and (6.99), we obtain an equation describing behavior of the studied material under hot-probe measurement conditions.

$$U_L = \frac{L}{q\mu n} j_s + \beta \Delta T \tag{6.103}$$

When a heater comes in contact with the electrode, electrons (in the case of the n-type semiconductor) run up into a semi-infinite space of a conductive matter. So, the current density will be related with the measured current, i_s, according to the following equation:

$$j_s = \frac{i_s}{2\pi r^2} \tag{6.104}$$

where r is a distance from the heated electrode ($2\pi r^2$ being the surface area of the hemisphere). Propagation of electrons in the space shaped in the form of a thin disc is limited by the disc thickness. Therefore, Eq. (6.104) transforms into the following approximated relation at the distance of integration:

$$j_s = \frac{i_s}{2\pi dL} \tag{6.105}$$

After substitution, Eq. (6.103) transforms in to the following final expression:

$$U_L = \frac{i_s}{2\pi q\mu nd} + \beta\Delta T \tag{6.106}$$

This equation, describing the system behavior in both excited and relaxed states for the wide-bandgap semiconductors, may be solved for the free charge carriers concentration:

$$n = \frac{i_s}{2\pi q\mu d(U_L - \beta\Delta T)} \tag{6.107}$$

Thus, we can calculate the concentration of charge carriers using the experimental data. In the case of low-bandgap semiconductors, we need to consider also the additional intrinsic charge carriers Δn, excited by the heating (see Eq. (6.96)).

The concentration of free charge carriers depends significantly on the thickness of the studied thin film (see Eq. (6.105)). Thus, the method and accuracy of thickness measurement will affect the results of our calculations. An error in determining the thickness leads to a deviation in the value of the concentration of charge carriers.

Moreover, it is easy to show that the experimental error in determining the concentration of free charge carriers will always be less than the experimental error in measuring the thickness:

$$\frac{\Delta n}{n} = -\frac{\Delta d}{(d + \Delta d)} \tag{6.108}$$

Therefore, the thickness becomes a decisive factor in determining the accuracy of the thin film parameters extracted from the hot-probe characteristics.

A real experiment is a process driven by both variable controlled parameters and random variables that influence the expected results. In the case of dynamic hot-probe measurement, these casual random values can represent a small deviation of the temperature of the heated electrode from the specified one, a change in the contact of this electrode with the sample being measured, etc. Therefore, in order to reduce the influence of various random parameters on the shape of the recorded hot-probe characteristics, it is desirable to carry out several measurements under the same conditions. Recorded characteristics in tabular form should be averaged to use in the following calculations. Obviously, the measurement accuracy increases with the number of measurements. Our experience suggests that in order to achieve the required process accuracy, it is sufficient to carry out three different measurements at the same temperature and process duration in different places of the studied thin film.

Figure 6.44 represents the laboratory home-made hot-probe station. It consists of the XYZ table for a sample arrangement, two spring-loaded electrodes and a heater enabling to keep the required temperature on one of the electrodes. The measuring electrodes are connected with the digital multi-meter that measures the voltage between electrodes or the current between them and transfers measured data in the computer. Obtained data in the table or graphical form may be presented and used for estimation of the basic parameters of the thin films or bulk samples.

The principle of operation of this technique is that the heated majority charge carriers, electrons or holes, have a higher velocity of

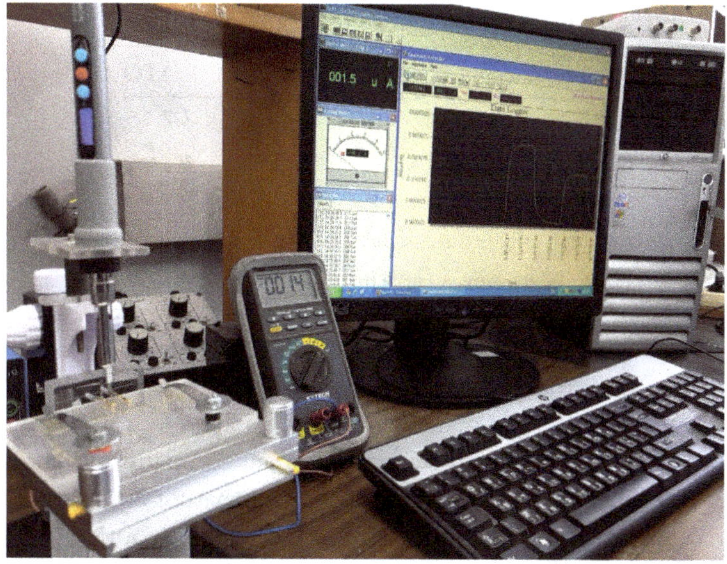

Fig. 6.44 The laboratory home-made hot-probe measurement setup.

movement than the cold ones. As soon as we begin to heat one of two electrodes, joining the surface of the semiconductor, the major charge carriers run away from it. In this way, a potential difference arises between the hot and cold electrodes, the sign of which is determined by the type of major charge carriers. If the heated electrode is connected to the positive terminal of the measuring voltmeter, electronic conductivity will show a positive voltage and a hole-type material will show a negative voltage. If we short-circuit the hot and the cold electrodes with an ammeter, a current will begin to flow through it. The sign of the current is defined by the type of the major charge carriers. Analysis of the measured dynamical characteristics of voltage and current dependencies on the temperature conditions, allows us to determine the recombination rate, the concentration and the mobility of majority of the carriers. Also, the magnitude of the potential difference, measured at different temperatures, enables us to determine the Seebeck coefficient (a thermopower) of the material.

Let us consider the properties of thin indium-tin oxide (ITO) film deposited on the glass slide with dimensions 25×75 mm^2 and

Table 6.1 Basic properties of ITO thin films.

Parameter	**ITO**
Energy gap, Eg (eV)	3.7
Relative permittivity, $\varepsilon_{r,\infty}$	4
Static dielectric constant, $\varepsilon_{r,0}$	9
Electron effective mass, m_n^* (m_0)	0.3
Hole effective mass, m_p^* (m_0)	0.22
Typical electron concentration, n_c (cm^{-3})	~5×10^{20}
Mobility of electrons, μ_n ($cm^2v^{-1}s^{-1}$)	26

thickness of 1 mm prepared by a commercial company and characterized in our laboratory. ITO represents a transparent conductive coating that has many applications in electronics and optics. This is a wide-bandgap semiconductor of the degenerate type concluding high concentration of majority carriers. Its properties significantly depend on the growth conditions. Thus, the indium oxide, In_2O_3, has a bandgap of ~3.7 eV [7]. To obtain enough high conductivity while retaining high transparency, this material is applied with alloyed tin (Sn) in relation 5–10%. Sn produces interstitial bonds with oxygen inside the In_2O_3 structure by replacing the In^{3+} atoms on the Sn^{2+} and Sn^{4+} which exist either as SnO or SnO_2 states, respectively. However, a predominance of the higher valency in Sn acts as a n-type donor releasing electrons to the conduction band. Thus, in ITO, both interstitial tin and oxygen vacancies contribute to high conductivity. The basic average properties of ITO are presented in Table 6.1.

The sheet resistance of the studied ITO sample measured using four-point probe method was of $R_\# = 8.3\ \Omega/sq$ that correlates with the data of the manufacturer (10 Ω/sq). The thickness of the ITO film calculated by the transmittance characteristic was of 150 ± 10 nm or ~150 nm. Recorded dynamical hot-probe characteristics for the ITO sample are shown in Fig. 6.45.

Figure 6.45(a) illustrates the dependence of a hot-electrode voltage on the processing time for different temperatures and Fig. 6.45(b) shows the current measured between hot and cold

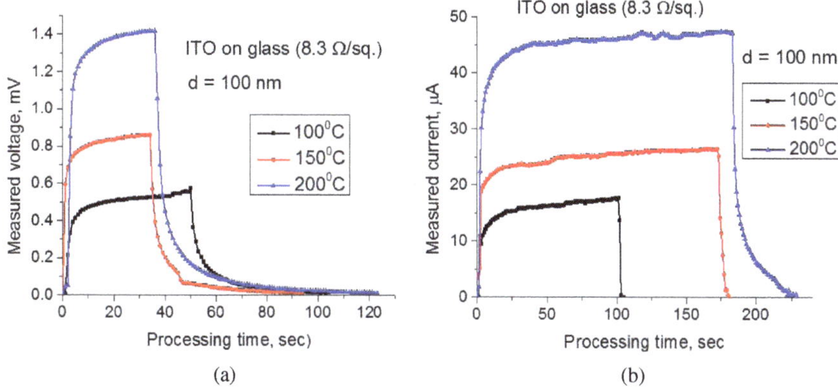

Fig. 6.45 Dynamical hot-probe characteristics of the commercial ITO coating: (a) a voltage measured between hot and cold electrodes; (b) a current between electrodes.

electrodes, measured for the same temperatures. The curves shown in Fig. 6.45(a) represent a positive voltage measured between hot and cold electrodes for different temperatures. This shows that the ITO coating is a semiconductor of n-type. As shown, the dynamical curves repeat each other, therefore, all dynamical processes in the film are the same for different temperatures. In the recorded plots, three different areas can be identified: the region of a steep increase in voltage due to heating, the region of steady state voltage and the region of a sharp decrease in voltage when the heater is disconnected from the hot electrode. All these processes occur in the absence of some external electrical or magnetic fields, so, they are driven by the temperature difference only. The steep rise of the voltage between electrodes and the current reflects a run up of hot electrons from the heated electrode. This rise happens very quickly due to differences between rates of the heated and cold electrons. Evidently, these rates are proportional to the root square of the heating temperature. Therefore, a higher charge difference will be obtained for higher temperature, which is confirmed by measurements.

Measured values of the voltage decreasing after removing a heater from the hot-electrode are presented in Table 6.2. This table also

Table 6.2 Measured voltage and basic calculated parameters of the ITO film.

V(t) T^0,K	Equation	T_0 = 373 K	T_1 = 423 K	T_2 = 473 K
$V(t_0)$		0.576	0.865	1.420
$V(t_1)$		0.368	0.573	1.060
$V(t_2)$		0.288	0.397	0.703
y_0'	6.102	−0.272	−0.349	−0.362
τ(s)	6.109	2.12	2.48	3.92
μ(cm²/V · s)	6.101	14.74	10.90	6.22
β (μV/K)	6.102	3.12	8.44	13.76

contains a calculation of several parameters, performed using the measured data. By definition, the relaxation time may be calculated using a derivative in the first point of the diagram (see Eq. (6.89)):

$$f(t)=f'(t_0)\tau+f(t_i);\ \tau=-\frac{f(t)-f(t_i)}{f'(t_0)} \tag{6.109}$$

Here, $f(t_i)$ is the initial value of the measured parameter, a voltage. Now, using the three-point formula (6.102) and the relation (6.109), we can calculate the relaxation time for all three diagrams as shown in Fig. 6.46.

The thermopower of the ITO film can be calculated using Fig. 6.47, which shows the dependence of the potential difference on the temperature of the heated electrode.

The results obtained for the mobilities show that they obey the formula (6.39). So, the main relaxation process in industrial ITO films is lattice scattering or interaction of excited electrons with phonons.

The calculation of the number of charge carriers moving under thermal excitation can be performed using Eq. (6.107) applied to the first (ascending) part of the hot-probe characteristic. Figure 6.48

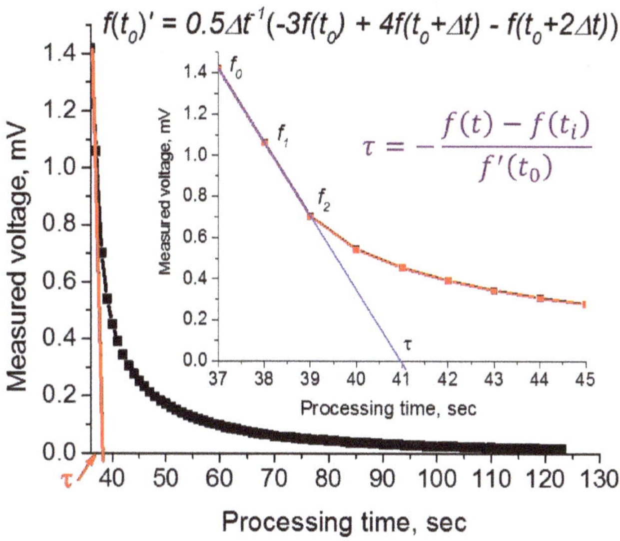

Fig. 6.46 Derivation the relaxation time from hot-probe measurement; inset shows the characteristic processing.

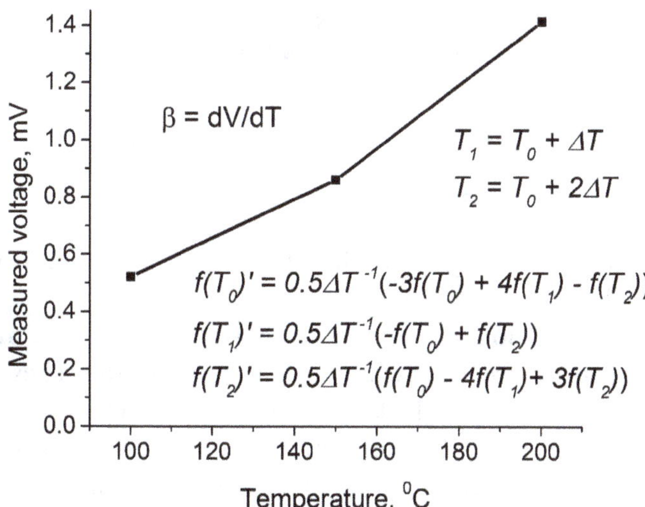

Fig. 6.47 Dependence of the potential difference on the temperature.

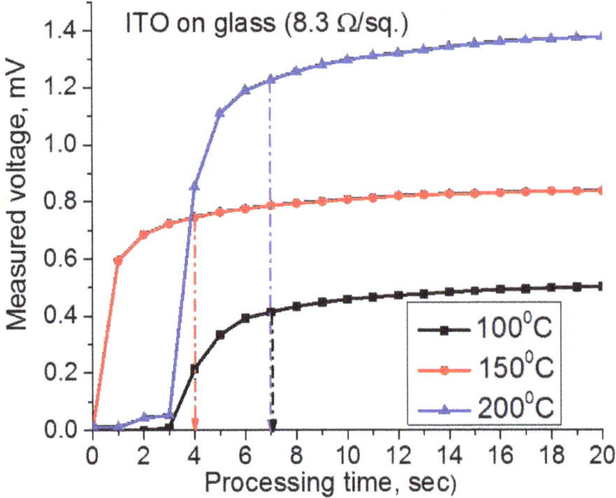

Fig. 6.48 Extraction of saturation parameters, voltage and current, from recorded dynamical hot-probe characteristics.

Table 6.3 Measured voltage and current and calculated parameters of the ITO film.

Temperature, T (°C)	**Rise Time, Δt (s)**	**Measured Saturation Current, i_s (μA)**	**Measured Saturation Voltage, U_L (mV)**	**$n \cdot 10^{20}$ (cm^{-3}) Eq. (6.107)**	**σ (Ω · cm)$^{-1}$ Eq. (6.99)**
100	4	11.5	0.414	2.64	$0.63 \cdot 10^3$
150	4	19.6	0.747	3.67	$0.64 \cdot 10^3$
200	4	35.2	1.227	6.93	$0.69 \cdot 10^3$

illustrates the processing of the first part of the recorded hot-probe characteristics.

For the calculation, we use the steady-state values of the measured voltage and current reached within a certain short time. These parameters, together with the calculated mobility and thermopower, make it possible to calculate the concentration of charge carriers and the conductivity of the coating by the formula (6.99). Calculation results are presented in Table 6.3.

Comparison of the calculated resistivity of ITO films provided using the formula (6.99) with the calculation based on our measurements, $\rho = R_{\#} \cdot d = 0.83 \cdot 1.5.10^{-4} = 1.25 \cdot 10^{-4}\ \Omega \cdot$ cm, shows good

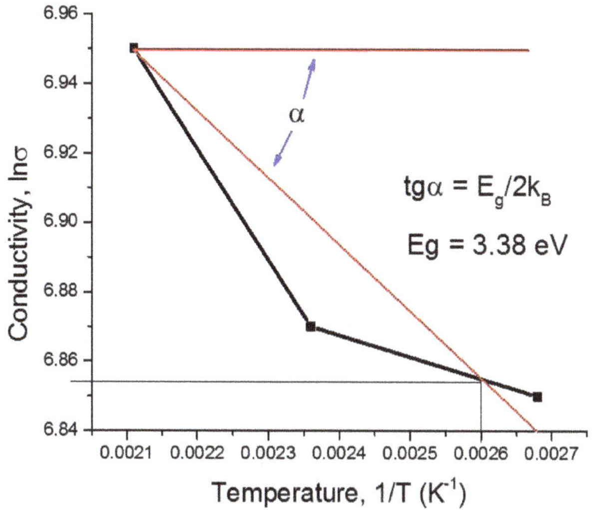

Fig. 6.49 Evaluation of the semiconductor bandgap by experimentally obtained conductivity.

coincidence. Using calculated conductivity from Table 6.3 and Arrhenius's relation, we can estimate the bandgap of the ITO layer as shown in Fig. 6.49 (see also Eq. (6.48)):

$$\sigma_T = \sigma_0 e^{\frac{E_g}{2k_B T}}; \; E_g = 2k_B \frac{\ln\sigma_1 - \ln\sigma_3}{T_3^{-1} - T_1^{-1}} \quad (6.110)$$

Bandgap of the commercial ITO film calculated by formula (6.110) is equal to 3.38 eV. Calculated parameters obtained from the recorded dynamical hot-probe characteristics of commercial ITO films are in agreement with the reference data. The measurement and recording technique is simple and low-cost.

Bibliography

Hudson, Derek J., Statistics, Lectures on Elementary Statistics and Probability, Geneva 1964.

Box, George E.P., J.S. Hunter and W.G. Hunter, Statistics for Experimenters, Wiley-Interscience, 2nd Ed., 2005.

Napolitano J., Experimental Physics, Notes for Course, Rensselaer Polytechnic Institute, 1999.

Fisher R.A., The Design of Experiments, Oliver and Boyd, Edinburgh 1935.

Draper N.R., H. Smith, Applied Regression Analysis, John Wiley & Sons, 1981.

Venables John A., Introduction to Surface and Thin Film Processes, University Press, Cambridge, 2003.

Seshan Krishna, Handbook of Thin-Film Deposition Processes and Techniques, 2nd Ed., Noyes Publications, 2002.

Pierson H.O., Handbook of Chemical Vapor Deposition, 2nd Ed., Noyes Publications, 1999.

Ohring Milton, Materials Science of Thin Films, Deposition and Structure, 2nd Ed., Academic Press, 2002.

Pranevicius L., Plasma Technologies, Vitautas Magnus University, 2003.

Wasa K., M. Kitabatake, H. Adachi, Thin Fim Materials Technology, Sputtering of Compound Materials, Springer, 2004.

Umrath W., Fundamentals of Vacuum Technology, Oerlikon Leybold Vacuum, Cologne, 2007.

Hass G. and R.E. Thun, Physics of Thin Films, Vol. 3, Academic Press, 1966.

Sze S.M., Physics of Semiconductor Devices, 2nd Ed., Wiley-Interscience, 1981.

Tumanski S., Principle of Electrical Measurement, Taylor & Francis, 2006.

Maissel L.I. and R. Glang, Handbook of Thin Film Technology, McGraw-Hill, 1970.

Schroder D.K., Semiconductor Material and Device Characterization, John Wiley & Sons, 1998.

Schmidt W., Optical Spectroscopy in Chemistry and Life Sciences, Wiley-VCH, 2005.

Van Zeghbroeck, Bart J., "Principles of Semiconductor Devices", University of Colorado at Baulder, 1999, See also: http://ece-www.colorado.edu/~bart/book/

Index